# Decantations

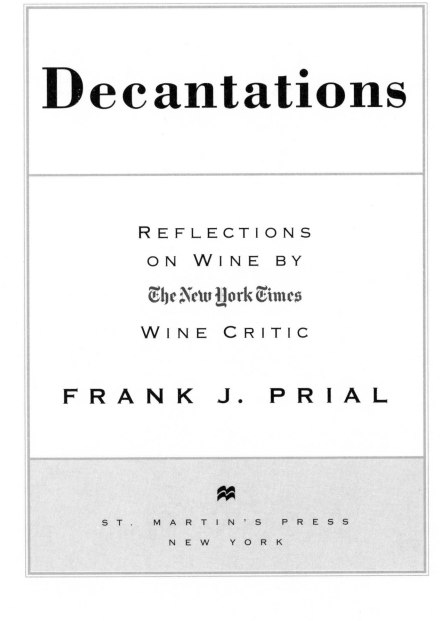

# Decantations

REFLECTIONS
ON WINE BY
The New York Times
WINE CRITIC

# FRANK J. PRIAL

ST. MARTIN'S PRESS
NEW YORK

www.stmartins.com

BOOK DESIGN BY DEBORAH KERNER

ISBN 0-312-28443-8

First Edition: November 2001

10  9  8  7  6  5  4  3  2  1

To Grace, Joseph, James, and Hazel

AND TO OTHERS

WHO MIGHT ONE DAY

JOIN THEIR MAGIC CIRCLE

# ACKNOWLEDGMENTS

Bylines, in books as in newspapers, give the impression that the writer did all the work. Don't believe it. A newspaper is a collegial endeavor; so is a book. Not a single byline, certainly not the one over *The New York Times* wine column every week, would appear without the help of a skilled support staff, from clerks, photographers, and researchers to (in my case) editors who have saved my neck more often than I care to remember. I'm grateful to all of them. I particularly want to single out *The Times's* Mike Levitas, the man who makes authors out of reporters, for his unflagging conviction that there was a book here to be done, his persistence in finding a good publisher to do it, his counsel, and his Job-like patience with a foot-dragging writer. Thanks, too, to Diane Higgins, the book's editor, for wresting order and coherence from journalistic chaos, and to Katherine G. Ness for her superb job of copy editing a lengthy, accent- and umlaut-laden manuscript.

# CONTENTS

## 3. ALL IN THE FAMILY

## 4. THE SOMETIMES NASTY, SOMETIMES DELIGHTFUL, BUT ALWAYS INTERESTING BUSINESS OF WINE

## 5. PERSONALITIES

# INTRODUCTION

The first Wine Talk column appeared in *The New York Times* on July 7, 1972, almost thirty years before the publication of this new collection. It was a modest debut: no advance notice, no fanfare, purposely low-key. With good reason. No one was sure what the column should be like or even how long it would last. There had been wine stories in *The Times,* on the food pages and in the business section, but never on a regular basis. Would there be enough to write about? Public interest in wine was hardly overwhelming in the early 1970s. But *The Times* was broadening its "life-styles" coverage just then, and wine seemed to be a logical addition to the mix. Besides, other major newspapers were running wine items and a few already published wine columns. Also, there were more than a few potential wine advertisers wondering when *The Times* would get involved.

I had joined the newspaper in 1970 with no thought of writing about wine. But I was curious enough about the subject to do a story, early in my tenure, on Nicolas, the big French wine company, while on vacation in Paris. At the behest of Charlotte Curtis, then head of the Living Section, I did several more wine stories. Shortly thereafter, when the idea of a column was broached, I was tapped to do it. Who could say no to that? "We'll try it for a couple of months," said Abe Rosenthal, then *The Times*'s managing editor. A couple of months became almost thirty years. But not uninterrupted.

Originally the wine column was a sideline, what the French call a *violon d'Ingres.* In 1979 I dropped the column and went overseas as a foreign correspondent; I picked it up again in 1984. In 1986 Max Frankel, the new executive editor, sprang me for another couple of years to try other things. Both times, however, I kept my hand in, continuing a twice-monthly column on wine in the Sunday magazine. That effort, begun in the mid-1970s, flourished until 1992. Over time, as food—and drink—assumed a larger and larger role in *The Times,* it became evident that wine was no longer a story that could be covered from a desk on West 43rd Street, the paper's headquarters. Originally relegated to the back pages of the Saturday edition, Wine Talk was moved in 1977 to the new, greatly expanded Wednesday Living Section. Wine was no longer a journalistic afterthought; it had become a legitimate news beat. Columns and other wine stories from France, Italy, Germany, Spain, Australia, and South Africa began to appear.

California, relatively insignificant as a source of fine wine when the column began, soon provided a steady stream of good stories. It's difficult to believe now that, early on, some readers complained that we overemphasized the vibrant, fast-growing California wine industry. It was a flash in the pan, they said, some suggesting that I must be in the pay of the Californians to write so much about them. Covering California's emergence as one of the great wine regions of the world has been great fun, although I miss the days when August Sebastiani, one of the shrewdest men in the business, dressed up in bib overalls and wandered around his tasting room pretending to be a simple farmer. I miss being able to knock on the door at the Mondavi Winery and have it opened by Bob Mondavi—who never pretended to be a simple country fellow. In the early days, there were cafés in the California wine country and the people frequenting them grew grapes or made wine. Now there are bistros and the people in them have titles—director of this or coordinator of that—but no juice stains on their shirts or dirt under their fingernails. It's the price of success. (In fact, the farmers and the cellar rats are still around; they're just harder to find.)

Europe has changed as well, where wine is concerned. There, too, most of the old lions are long gone: Philippe de Rothschild, a marketing genius who was also a poet; Alexis Lichine, who showed the Bordelais how to get on with the rest of the world; Ignacio Domecq, "The Nose," whose lifestyle probably outshone that of Louis XIV; and André Gagey, one of the great gentlemen of Burgundy. Once Bordeaux and Tuscany seemed remote and exotic. Now, with faxes and e-mail, European winemakers can read about themselves in Wine Talk before most people in New York, including Wine Talk's author, have gone to bed. In the farthest corners of western Spain or northern Sardinia it's commonplace to find winemakers just back from a year in Sonoma County or Australia. Older folk who have never left the village ask me with awe: "Do you know Robert Parker?" (I do and they're right; he's a pretty awesome fellow.)

Many of the stories in this collection date from the years when, along with wine, I covered what we in the media persistently—and inaccurately—refer to as straight news. I like to think many of those pieces benefited from a kind of synergy. Sometimes I would hear things like "How can a wine writer cover the United Nations Security Council?" Just as often, I would be fed some interesting tidbit from a member of the Security Council who happened to love wine. Federal judges, and an occasional politician, proved helpful from time to time simply because they were interested in wine. Once, a story of mine about a NATO meeting in Brussels appeared in the same edition of *The Times* as a column by me about Beaujolais. A *New Yorker* magazine humorist found the juxtaposition of the two bylines bizarre enough to warrant a piece in that magazine. It didn't come off, but her interest was appreciated.

The period covered by the stories assembled here, mostly in the 1980s and 1990s (a previous collection dealt with the earliest years), mark America's establishment as a wine-drinking nation. Frankly, I didn't think, twenty years ago, that it would ever happen. We consume far less than most of the wine-drinking countries, including France, Italy, and even Argentina and Russia: two gallons per capita per year compared with up to seventeen gallons. We may never reach those levels. Our wine consumption has leveled off in recent years, and most of the country's wine intake is accounted for by a surprisingly small number of people. Even so, no one thinks any longer that wine is just a passing fad. A couple of days in Las Vegas, watching the macho gamblers drinking great Bordeaux and California cabernets with their steaks, should be all it takes to convince anyone that wine is here to stay.

Almost from the start of Wine Talk, I resolved to downplay the technical side of wine, the minutiae of grape growing and winemaking, along with the long lists of tasting notes that are the lifeblood of the wine newsletters and specialty magazines. In part it's a problem of space, a problem partially solved in recent years by sharing that section of the report with my colleague Eric Asimov. Beyond that, it's probably safe to say that only a handful of the more than 1 million people who read *The New York Times* every day are seriously interested in wine. Like every other part of the paper, the wine column should offer something that all or most of its readers can enjoy or at least find informative. You don't have to be a baseball fanatic to read a story about a great player or a particularly exciting game. You should not have to be a budding enologist to enjoy reading about wine.

The story of wine, from its beginnings some six thousand years ago to this year's Beaujolais nouveau, is part of our social and cultural history. It's an ongoing saga, filled with exciting events and extraordinary people, and it is those events and those people who over the years have mattered most in Wine Talk. In truth, going the *ad hominem* route, concentrating on people rather than things, may not have been a conscious decision after all. The best stories in any newspaper are about people. That's what we do best. Most reporters, even those disguised as wine columnists, wouldn't have it any other way.

Most of these columns are of fairly recent vintage. But a few go back twenty years. I thought of updating them; some of the people are gone, some of the prices—and quite a few of my opinions—are sadly out of date. But I decided no, better to let them represent the period in which they were written. The 1995 Bordeaux were very good, but so were the 1982s.

# Decantations

# CHAPTER 1

# WINE WRITINGS
# AND WRITERS

For whatever reason, there is very little humor in wine writing. Why, I wonder. So much has been said about how wine brings joy to so many lives, how it lightens one's burdens, encourages sociability, brings roses to the cheeks and lightness to one's step. It's not about death and taxes. So why is this happy message so often conveyed in somber words?

In part, it's the result of our language, which was still English the last time I checked. The first modern wine writers were English. They saw wine as exotic because that's the way they saw the countries that produced it—France, Portugal, Spain, and later Italy. We inherited their romantic style of wine writing and, sad to say, adopted it uncritically: the worshipful tone, the classical allusions, the elitist judgments.

We have a long way to go but we are beginning to relax. As well we should—we Americans are winemakers, after all, and our British cousins, with all their enological enthusiasm, are not.

Our contemporary humorists—Russell Baker, Calvin Trillin, Dave Barry, Art Buchwald—have all taken well-aimed swipes at the wine game from time to time, but none of them has equaled A. J. Liebling. One of our best observers of the wine scene is an insider: Randall

Grahm. He is a wine writer; he is also a winemaker, and he will pop up more than once in these pages.

Too often, wine is still treated as something just short of miraculous and the people who make it as miracle workers. Very few journalists have been sharply and consistently critical of the wine trade, at least until recently. Pierre-Marie Doutrélant was one of the first. You will find a few of his slings and arrows herein.

## WORDS, WORDS, WORDS

Every so often, it is meet and proper to once again examine a peculiar subgenre of the English language—and of the American language as well—that has flowered wildly in recent years, like some pulpy jungle plant. It's called *winespeak*.

Winespeak is a branch (tendril?) of the mother tongue that seeks to render the sensory experiences triggered by wine into comprehensible words. You know—to explain what it tastes like. Winespeak leans heavily on metaphor. Or is it analogy?

A wine esthete pokes his sensitive beak into a glass of Champagne. Then, head thrown back, eyes closed, and visage wreathed in an otherworldly smile, he incants: "I see . . . I see . . . yes, I see a young girl in white, barefooted, running across a vast green lawn, long hair flowing." That's metaphor. Another initiate sniffs a white wine of questionable provenance. "Broccoli," he says, and tosses it out. That's analogy.

Winespeak—modern winespeak—can be traced to the Gothic piles of Oxbridge, where, in the nineteenth century, certain dons, addled by claret, bested one another in fulsome tributes to the grape. There is, of course, a much older winespeak. Here from the Old Testament is Joel 1:5, commenting, one suspects, either on an early closing law or on a shortage of the nouveau Beaujolais of his time, when he cries, "Awake, ye drunkards, and weep; and howl, all ye drinkers of wine, because of the new wine; for it is cut off from your mouth." Maybe it's better in Aramaic.

The Oxbridge worthies, steeped in Homer and Tacitus as well as Madeira, leaned on antiquity for their prose. It would be unfair to say that Bacchus and Dionysus were the Bartles & Jaymes of Magdalen and Christ

Church in Victorian days, but I think you get my drift. Wines were forever reaching Parnassian, if not Olympian, heights, and a third-rate Médoc could transport even a red-brick lecturer in Greek to the Elysian Fields.

We Americans, not entirely convinced that anything happened before 1930, have little of either history or tradition on which to draw to describe a wine. There is also, of course, an argument to be made that it's inappropriate to summon up some lecherous Roman deity to illustrate a wine made by a twenty-year-old Californian in a place where lately lettuce grew.

So we're not erudite. We're inventive, which is just as good. When a situation cries out for purple prose, we are not found wanting. Well, not usually. My friend Roger Yaseen, who is in investment banking but knows a bizarre turn of phrase when he sees it, offers the wine list from a restaurant called—sorry—Anotherthyme, in Durham, North Carolina. It's an okay list; someone there knows wine. But oh, the descriptions.

A 1979 Pol Roger is $35, and at Anotherthyme, this is what you're going to get for your money: "Frail lilies blessed with the permanence of granite. Fabulous, timeless vintage Champagne." "Timeless vintage" is probably an oxymoron, but after the lilies and the granite, oxymorons just don't seem to matter.

The next Champagne, Dom Ruinart, has "exceptionally bright flavors and a stimulating spritz," and drinking it, we are told, is "like pouring diamonds into a tulip."

Krug's Grande Cuvée has "mysticism, erudition and taste," but what are they compared to Schramsberg's 1981 Blanc de Noirs, a "glorious flesh-colored fluid."

When it comes to simple white wines at Anotherthyme, I couldn't decide between Château Guiraud's "G," "like cool wet sand under pearly seaside light," or a Montagny Les Coeres from "the scrubby hillsides south of Beaune . . . whose lithe powerful body is insulated by soft tender curves."

Tender curves are hard to pass up, but how could I ignore the Piersporter Hoffbürger Spätlese with its "smirking beauty of fruit without pride." No, really; that's what it says.

A Fleurie—which is, after all, just a high-class Beaujolais—has "a texture like the sinful strokes of a feather boa," and Chiroubles, another Beaujolais, is "flashy Chiroubles, adored by café society Parisians, with Beaujolais' renowned soft juiciness and a provocative nip."

There is a white Graves, La Louvière, "whose sound architecture aptly frames its core of minerals which seems to have been mined from the bowels of the earth," and there is a Gevrey-Chambertin that is "fathoms deep and as pleasingly prickly as a kitten's tongue."

There is a Firestone cabernet "juicy and ripe with a flexible spine," a Franciscan merlot of "smooth soft body with ample underlying muscle," and a Kalin pinot noir with "a wooly welcoming feel." Along the way, one gets the feeling that Ben Jonson or John Marston, or even Gerard Manley Hopkins 250 years later, would have loved this crazy imagery. After all, wasn't it just a few years ago that some California fellow, actually a screenwriter seeking an honest dollar, became immortal by describing some clunky big zinfandel as "a Roller Derby in the mouth"?

When it comes to adjectives, I prefer something a little drier myself. Still, I'd love to eat at Anotherthyme one day. I wouldn't be at all surprised to find Alice, the Mad Hatter, and the Rabbit, or maybe Jonathan Winters, at the next table. And *not* drinking tea.

MARCH, 1987

## SEARCHING FOR THE RAREST VINTAGE: HUMOR

There's a grape shortage in California. That's unfortunate, but it will pass. Someone will grow more grapes. Then there will be too many. Happens every time.

But there is a more serious shortage in the wine world, one that doesn't ever seem to go away. Not of grapes or vines or oak barrels or winemakers or even customers. It's humor.

Not the back-slapping, hail-fellow-well-met stuff; there's always plenty of that around. I'm talking about the real thing: the sly dig, the witty aside, the mordant observation, the risible—yes, even the ribald. For whatever reasons, wine just doesn't seem to evoke the comic muse.

Not directly, anyway. What humor there is, is mostly inadvertent.

Tasting notes, for example, can be hilarious. "Firm, chunky and herbal, with a range of aromas that includes forest underbrush." "Fig and pear notes that stay with you." "Compact and tightly wound." "Impressive for its focus." "Jammy and lush until the muscular tannins kick in." "Lean citrus notes in a tightly wrapped package." Wines are "fat" and "chewy" and occasionally "taut." They have "gobs of fruit" and "aromas that leap from the glass."

People who make wine and people who drink it understand the immense difficulty of putting sensual experiences into words. To an outsider, however, the results can be sidesplitting. The art world has the same problem. Of course, art critics can always fall back on clever sarcasm. Wine critics generally ignore what they don't like.

Wine catalogues occasionally flash a bit of humor—again, without knowing it. When an advertisement calmly announces that a bottle of 1993 Burgundy sells for $900 or a 1990 Bordeaux is $400, one has to decide whether to laugh or cry. To laugh is to see something truly appalling in a humorous light. Gallows humor, of course.

In recent years, catalogues have branched out. They now hawk everything from bizarre bottle openers to burgundy-colored beach blankets. They sell wine neckties, wine T-shirts, and wine jewelry. They are, in fact, wine kitsch, and good kitsch is always funny.

Somewhere along the line, desktop publishing and the public relations trade combined forces to invent winery newsletters. They arrive in a steady stream, four relentlessly cheery pages of anecdotes about the winery cat and the new tasting room, some thoughts on the weather, and snapshots of the whole gang gathered around the crusher-stemmer and brandishing glasses of merlot.

Warmhearted? Yes. Humorous? Well, not exactly.

And then there is Randall Grahm. Mr. Grahm runs a vineyard and winery called Bonny Doon in Santa Cruz, California. There has long been the suspicion that the winery is only an excuse for putting out a newsletter. Mr. Grahm makes wines with names like Big House Red, Old Telegram, and Le Cigare Volant, which should give some idea of the verbal high jinks he perpetrates in his newsletter.

To describe Mr. Grahm's style as stream of consciousness would not be inaccurate so long as it's clear that his stream is like that of a high-powered winery hose. It's a riotous jumble of gags, impenetrable puns, wine cellar technobabble, and other assorted *jeux de mots,* most of them impossible to excerpt.

He writes: "I schlepped (albeit not languidly, doc) through southern France . . ."

Going after the heavy-handed writers of tasting notes, he describes his own Cigare Volant, which means "flying cigar" or "flying saucer," as "a wine for the urban hunter/gatherer."

"It is," he goes on, "like living to be 200 years old. It is a bouquet of ultra-violets. It is the sun pouring through one's sievelike body. It is the taste of the colors of mauve, nutmeg and rosemary, the musted moan of violaceous velvet. It is all of the virtues and more vices than are dreamt of in Miami." And so on.

His wine list (he does make his living selling the stuff) includes half a dozen 1994s, some 1993s, a couple of 1995s, and ends up with a 1972, a Citroën DS-21 station wagon that he has been trying to unload for several years. Price $9,000, reduced from $10,000.

Sadly, the last place to look for even the thinnest strains of wine wit is on a restaurant wine list. This is supposed to be serious stuff. And watching a nervous guest trying to decipher a list makes it clear just how serious it is.

It is the perfect moment for a couple of laughs and now, finally, someone is providing them. On the list at Layla, Drew Nieporent's Middle Eastern theme restaurant in New York, each wine category begins with a question; the wines that follow are supposed to be the answers.

*What to drink with baba gannouj and a mess of other mezze?*
Rosé wines; one each from Tuscany, Sicily, Provence, the Napa Valley, and the Rhône.
*What wines did Noah bring on the ark?*
Two white Burgundies and three California chardonnays.
*What is the cheapest red wine on the list?*
Las Campanas 1991, Navarra (Spain). $13.

(Just for the record, the cheapest white on the list is a $10 bottle from Crete called Kourtaki White, and the most expensive wine is Dom Pérignon 1988 at $185. There are seven wines for $15 or less.)

In keeping with the Middle Eastern theme, Layla offers wines from Lebanon, Crete, and mainland Greece. In fact, the Mediterranean possibilities are far from exhausted. Wines from Turkey, Israel, Tunisia, Algeria, and Morocco would make the list even more exotic.

One question on Layla's list asks "What were King Tut's top five reds?" They turn out to be from Spain, Portugal, Italy, and France. It would be appropriate to offer an Egyptian wine. Unfortunately, viticulture in Egypt has slipped rather badly over the last five thousand years. Better to stick with a nice Côtes du Rhône–Villages.

MAY, 1996

## A SAD EXAMPLE

January is the time of year when even normally prudent people lose all sense of caution and succumb to the urge to predict things. For futurologists, amateur and professional, who have had some measure of success in the past, it's a particularly dangerous time; they are expected to repeat their accomplishments yet again.

For those who have never had any luck predicting anything, the task is much less burdensome. What the heck—they have nothing to lose. Accordingly, we hereby scramble far out on a limb to disclose some of the more important wine events that will almost certainly come to pass over the next twelve months. The dates, it must be said, are only approximate.

JANUARY 26: Four more bottles of wine bearing Thomas Jefferson's initials will mysteriously turn up at a wine auction in Zurich. No one will profess to know whence they came. Collectors will accept the three bottles of Château Lafite 1776 as authentic but will voice considerable reservation about a single bottle of nonvintage Ridge zinfandel. They will point out that Jefferson's diaries frequently alluded to his strong distaste for the variety.

FEBRUARY 3: The new owner of one of the 1776 Lafites will display it to some friends. When they question his probity, he will hit one of them with the bottle and smash it. An insurance company will refuse to pay off, noting that Jefferson bottles have been turning up every few weeks.

MARCH 14: A band of masked men will board the controversial Napa Valley Wine Train just as it passes Mustard's Restaurant, north of Yountville, California. After frightening some elderly tourists from Iowa who think they are headed for Disneyland, the men will invade the galley and turn up the heat under the *pintadeau désossé,* simply ruining lunch for everyone.

Witnesses will later testify that some of the masked men sounded like winery owners who are opposed to the operation of the train, but charges will be dismissed for lack of evidence.

APRIL 7: In Paris, LVMH, the Moët Hennessy–Louis Vuitton conglomerate, will announce that it has acquired chateaus Lafite-Rothschild and Mouton-Rothschild. The announcement will say that the two will be combined into a new entity known as Château LRMR. A spokesman will say that the new subsidiary will embark on an aggressive, vertically integrated acquisition program that will provide new outlets for its wine. Included will be fast-food chains, international vacation clubs, and cruise lines.

APRIL 8: In Tokyo, Mitsubishi will announce that it has acquired LVMH.

MAY 15: Robert Mondavi will disclose that he has been doing a little home winemaking since his retirement "to pass the time." This year's production, just for family and friends, will be about 80,000 cases, an aide will report.

JUNE 12: A six-foot-tall, three-hundred-gallon bottle of cabernet sauvignon will top the bidding at the annual Napa Valley auction. It will be knocked down for $300,000 because it's such good wine and because it comes wrapped in its own BMW mini-van. The story will get immense play on the food pages and on televison and will convince approximately 1 million potential wine drinkers to stick to beer.

JULY 20: The Environmental Protection Agency will announce that it holds to its earlier conclusion that the amount of the fungicide procymidone in some imported wines is not harmful. But the agency will say that it has decided to continue its tests for a year or so. "We'll keep everybody posted," a spokesman will say.

Meanwhile, to mollify a few strident critics, the agency will release some 1988 Beaujolais nouveau that had been withheld from the market for its tests. A prominent New York retailer will immediately take out full-page newspaper ads announcing a 1988–91 twin pack: "Two nouveaux practically for the price of one."

AUGUST 18: The Domaine de la Romanée-Conti will announce—with deep regret—that the 1990 Romanée-Conti will cost $5,000 a bottle. Mme. Bize-Leroy, an owner of the D.R.C., will acknowledge that the price does not exactly reflect production costs but will explain that the Domaine always tries to position itself slightly behind Château Pétrus.

AUGUST 19: In Tokyo, Mitsubishi will disclose that it has acquired Château Pétrus.

SEPTEMBER 15: Grape pickers in Bordeaux will tell reporters that because the summer was so hot and there was so little rain, the 1991 vintage should be quite good.

OCTOBER 15: Orley Ashenfelter, the Princeton University soothsayer, will announce that, using his econometric model and the Cray super-computer at the Institute for Advanced Study, as well as the information that the Bordeaux summer was hot and relatively rain-free, he has concluded that the 1991 vintage should be quite good.

NOVEMBER 2: The White House will propose an additional $6-a-bottle tax on table wine. Overriding industry objections, the president will point out that the tax will come to no more than a measly $1 a glass—using a four-ounce glass, of course.

NOVEMBER 4: Questioned on the ABC program *Nightline,* a White House aide will read from a budget report showing that using a two-ounce glass will bring the tax down to a minuscule fifty cents a drink. Irritated by Ted Koppel's aggressive questioning, he will characterize six-ounce and larger glasses as "holdovers from the self-indulgent Clinton years."

DECEMBER 5: The E. & J. Gallo Winery, the world's largest, will announce that it is moving from furnace-like Modesto, in the Central Valley of California, to leafy-green Sonoma County in an effort to enhance its image as a producer of fine wine. This plan could be changed next year if the legislature in Sacramento approves a long-standing Gallo proposal to move Sonoma County to Modesto.

DECEMBER 10: The Beaujolais impresario Georges Duboeuf, who in recent years has moved aggressively into the bottling and shipping of Rhône Valley wines, will announce that he has acquired the Rhône Valley.

DECEMBER 11: In Tokyo, Mitsubishi will announce it has acquired Georges Duboeuf.

JANUARY, 1991

## PEACHY KEEN

There are many ways to report on wine, but only two count. One is to gather lots of bottles, divide them into logical categories—for instance, red and white—then open them up and taste.

This is the proto-scientific, or clipboard-and-white-coat, approach. It requires cases of glasses and a serious mien, but it will produce sheets of notes that can be trimmed to fit any editorial space.

Much more pleasant is the "Hey, this isn't bad," the casual approach, in which one drinks around indiscriminately, then tries to remember later what the wines' names may have been. Neither approach prepares you for peach pasta.

Not so long ago, I enjoyed an undemanding little chenin blanc from a Napa Valley winery with the fey California name of Folie à Deux. Then I forgot about the wine—chenin blanc being eminently forgettable—until an article in a magazine called *Wine & Spirits* jolted my synapses.

In fact, it did more than that. Had not the name of the same chenin blanc caught my eye, I would never have discovered just how far we have come—pushed the envelope, as they say—in the new and uncharted business of matching food and wine.

Barbara Lang and Kris Ostlund, wine and food experts and an "exuberant young couple," according to the piece, were asked to create a menu around some Napa Valley wines that had been rated highly in another article in the same issue.

Along with the chenin blanc, they were asked to work with the Domaine Chandon Blanc de Noirs, a nonvintage sparkling wine; a William Hill 1987 chardonnay; a Christian Brothers 1985 cabernet sauvignon; a Robert Mondavi 1985 pinot noir; and a Franciscan Vineyards late-harvest 1986 Johannisberg Riesling.

Ultimately, most of their choices seemed reasonable enough and, judging from the pictures, fun to eat.

More intriguing by far was their employment of what, on first encounter, appears to be a new language to express a new discipline. A gastronomic newspeak, if you will, some of it lively and precise, a lot of it totally zany.

Early on, for example, to accompany the Chandon Blanc de Noirs, Ms. Lang and Mr. Ostlund concoct an appetizer of endive with shrimp and tarragon, blended with both cream cheese and sour cream. Different versions were tried and abandoned because, we are told, "Each version cut the fruit and finish out of the wine." Alas, the experts report: "The creamy-fat base simply wasn't wine-friendly."

Instead, they choose two other appetizers: prosciutto garnished with gremolada (minced dill and lemon rind) and chopped oysters with a sauce of sparkling wine, Champagne vinegar, tarragon, and shallots.

"The slightly salty, suave qualities of the prosciutto and gremolada boost the citrusy overtones of the sparkling wine," we are told, with a note that the Blanc de Noirs "seems to welcome these piquant and cleansing flavors."

Lamb is considered for the main course, and rejected. It's too heavy, too fatty, its flavor "simply too aggressive" for the wines. In a bold move, Ms. Lang and Mr. Ostlund decide to snap their collective fingers under the nose of the red-with-meat tradition and serve both the chardonnay and the cabernet with the main course. The two wines, they explain, "are at the very extremes of their varietal attributes—the chardonnay at the full, rich, opulent end; the cabernet at the lighter, fresher, fruitier, softer polarity." Their choice for the main course: a grilled loin of pork. "We chose pork as a friendly carrier for the two wines," the two explain. Of course, the pork demands a sauce. An orange-and-ginger composition proves "interesting in itself and with the pork, but distracting to both wines. It robbed them of their most beguiling qualities." A mustard sauce proved to be "delicious and zippy," but no match for the powerful chardonnay, "a big, toasty, oaky, buttery wine." The winner: a red-onion-and-fennel combination that permitted the chardonnay to remain "full and lush on the palate" and the cabernet to retain "its lovely fruit and soft roundness."

And so it went. Cow cheeses didn't make it with the "smoky, mint, spice, cherry flavors and soft tannins" of the pinot noir, but, happily, goat cheeses did.

The late-harvest riesling was married successfully—and bigamously, it would seem—to an apple tart with a maple cinnamon custard. The tart's "smoothness and delicacy," we are told, "intensifies the honey, spice and orange overtones of the wine."

Which brings us, out of sequence, back to the chenin blanc. Out of sequence, that is, unless you agree that the toughest challenges are best left to the last.

To Ms. Lang and Mr. Ostlund, the chenin blanc has—somewhat redundantly—a "melon, apple, peach-perfumed nose; fruit in the mouth, clean finish and a sweet impression of ripe fruit."

What would you serve with such a wine? They decide on pasta. Ah, but no ordinary pasta. No white clam sauce à la Bay Ridge is going to offend this wine. This pasta gets a peach sauce, a peach sauce "whose deep lush flavors could temper yet not totally obliterate a shock of sharp tarragon and zesty mint."

What sort of pasta could serve as "carrier" to this remarkable mélange? Fusilli and penne, serviceable old friends though they may be, are rejected. "Too heavy," our experts say. "We loved the sauce, but it cried out for a more delicate pasta." The cry was heard; thin spaghetti took the day.

"Contrasting the heady herbs with the sensuousness of the peach, the sauce then melding with the consistency of the cooked thin spaghetti turned out to be a superb backdrop for this chenin blanc," say the Ostlunds, for, yes, they are husband and wife. "Somehow," they continue, "the extraordinary texture of the peach sauce heightens the delicate impressions of the wine, making it creamier, even more appealing."

More appealing, indeed. But not just the food; the phrasing, too. Who knows? We may be breaking new ground here, reaching perhaps for the ultimate diet. We've all experienced dishes that look better than they taste; now we may have food and wine more fulfilling to read about than to eat and drink.

Put that on your peach pasta.

SEPTEMBER, 1989

## SHORT COURSE IN WINE TACTICS

It seems that everyone wants to learn something about wine these days. Not necessarily to drink it, but to learn about it. There are wine schools cropping up all around. Adult education centers offer wine courses; restaurants have them too. Even conservative universities, nervous about their dwindling enrollments, are coming up with wine courses. Tasting groups are everywhere, and wine books are appearing at a rate of one a month. Some fanatics go all the way and take up enology. Others get jobs in wineries. Some prefer hands-on experience, which they gather in a succession of bars.

Unfortunately, these methods all take time. They won't save you when you discover you're having dinner with some alleged wine experts. Most other monomaniacs—the tennis bore, the stock market bore, the hi-fi bore—can be handled. The wine expert is tough. After all, the stuff is inescapable; it's right there on the table in front of you, ready to be opened, perchance to be decanted, to be swirled, sniffed, and sipped until you're ready to go right through the wall.

Pleading ignorance is not only cowardly, it's bad tactics. To the committed wine bore, particularly if he is the host, the wine clod is a gift from heaven: a new audience, a possible convert. It's fire with fire, or nothing.

Here, then, is a short course in wine tactics. If you think it's going to get you into the Chevaliers de Tastevin, forget it. It won't even save you from being ripped off in your favorite liquor store. The sole purpose of the following information is to cover your ignorance in polite company.

The expertise offered here consists of a selection of simple words and phrases, all in the mother tongue. This meant eliminating some key terms in the wine expert's vocabulary, such as "Mon Dieu," "yech," and "feh"— but how much better to start off modestly.

The most important of these phrases is this: "It dies on the middle palate." Yes. Now repeat it six times. What does it mean? What difference what it means? Just say it when the supercilious host asks your opinion. With a bit of practice you can begin building sentences based on this phrase. For example, start with: "Superb, but—" Or you may wish to add "—but it finishes well."

This is the total wine put-down. Your host will have reverted back to sloe gin fizzes by ten A.M. the next day. After all, even Philippe de Rothschild can't argue with your middle palate.

Then there is the word "bramble." Do you know what a bramble is? It's a bush, right? Do you know what a bramble tastes like? Of course not— who eats bushes? Nevertheless, that's what you're going to say if the wine is red: "It has a real bramble taste; yes sir, a real bramble taste." Don't worry. It appears on a dozen different California wine labels, and it's a safe bet those guys don't know what it means either. For sure, your host doesn't.

To carry this kind of thing off you must dance and feint; never let 'em lay a glove on you. Don't start out calling a wine "oaky," even if you do have a vague idea of what you're talking about. The wine expert will hit you with "American oak or Yugoslavian oak?" He is piqued by now because the rest of the party has turned to you, impressed by your incomprehensible jargon.

More terminology: "cab," "zin" and "chard." These are three embarrassing little abbreviations much favored by California wine cultists. They stand for, of course, cabernet sauvignon, zinfandel, and chardonnay, and are employed thus: "We had a tasting of twenty zins and every one of them was really super." Or "He's an okay winemaker but there's too much oak in his cab and not enough in his chard." Most Californians can't stand "chard" either, but it does crop up now and then when they talk.

"Chewy" and "fat": These, you should be aware, are legitimate descriptive adjectives in some wine circles. They can be condoned—maybe—because more often than not they are used to convey some sense of wines that are truly indescribable: overpowering red wines that have too much of everything in them but restraint. The kind of wines one finds described as "big, fat, chewy monsters; can be laid down for decades but are perfectly good for drinking right now."

"Nose." As in "The nose is very forward." "Bouquet" and "aroma" are two different things where wine is concerned, but you need not concern yourself with them. "Nose" is a synonym for "smell"—and the only acceptable substitute. You can say the wine has a lovely nose or a peculiar nose or even a nonexistent nose. The polite way to say that the wine smells terrible is to remark that it has "an off-nose."

"Body." As in "This wine has excellent body." Note: Never say "This wine has an excellent body." That would be gauche, although someone once got away with—in fact became immortal by—saying of a wine that "it had narrow shoulders but very broad hips." "Body" simply refers to the substance of a wine.

Wine also has "legs." This is determined by swirling a partially filled glass of red wine and waiting for it to settle. If the glass is clear and clean and the wine any good at all, you should be able to see the lines of colorless glycerin still making their way down the inside of the glass. These are called "legs," obviously in keeping with our obsessive desire to equate wine with the human body.

The Germans are a bit more elegant on this one. They call the legs *Kirchenfenster*, or church windows, because as the glycerin comes down the sides of the glass it forms nearly perfect Gothic arches.

"Short." As in "The wine is pleasant enough but I find it a bit short." This is actually a useful term. It simply means that the taste of the wine does not linger in your mouth. This phenomenon is also referred to as a short finish. A wine whose taste lingers is said, naturally enough, to have a long finish.

You're now ready to handle "oak." As in "Thank God Mondavi is no longer obsessed with oak." Oak is the taste imparted to wine by the oak barrels in which it is sometimes stored. Enthusiasts argue over oak the way bears fight over territory, but at the moment oak is mostly out in wine circles. Unless, of course, it's subtle oak.

Oak, like short finish and long finish, is a tricky term. You'd better have some idea of what you're talking about when you use it. There is nothing a wine bully likes better than to be able to say to someone who has just pronounced a wine oaky than: "Sorry, but this one was fermented and aged in stainless steel. It's never seen wood."

"Alcohol." As in "What's the alcohol in this stuff?" This is an excellent phrase because it implies that you know what alcohol content means. A table wine that has close to 14 percent alcohol by volume is going to give you a headache if you drink too much, and it's probably too strong to go with your meal. But you don't have to know this. Merely posing the question will evoke some response from your host. All you need to do is to nod knowingly.

Also, "pH." As in "What's the pH in this stuff?" This term is even better than "alcohol" because nobody understands it. It has something to do with the intensity of the acid in a wine. Low pH means more intensity, high pH means less. Low would be around 2.85; high would be around 4.

These, then, are just a few of the words and phrases that wine people live by; words and phrases that you can master and use to your benefit without knowing muscatel from muscadet, or Romanée-Conti from Ripple. It's wrong, probably, to advocate such brazen chicanery but it will serve to give you breathing time if you choose to really learn something about wine. Also it will get you off unscathed if you choose never to mention or even think about wine again. And, too, as you begin to hear other people using these terms, you will come to realize how many of them know almost nothing about wine either.

JULY, 1983


## AFFAIRS TO REMEMBER

Wine drinking goes back at least six thousand years. Wine writing probably began a year or two later.

The Sumerians wrote about wine; so did the Babylonians and the Egyptians. Then came Homer and his wine-dark sea. In our own epoch, the

**15**

monks kept vineyard and vintage records at Cluny and the Clos de Vougeot, Pepys praised the "Ho-Bryan" he drank at the Pontac Head in London, and Jefferson, in Paris, reported on his trips to Bordeaux and Burgundy. But bookkeeping, diary entries, and ambassadors' reports are one thing. Celebrating wine as a cultural phenomenon is something else. The British seem to have invented it, along with Port and Sherry. Wine journalism is an even younger enterprise.

Which brings us to G. Selmer Fougner.

Never heard of him? Well, he was a journalist, and fame in that line of work is, as they say, fleeting. G. Selmer Fougner was probably New York's first newspaper wine writer. From the end of Prohibition in 1933 until his death in 1941, he wrote a daily—yes, daily—wine column in *The New York Sun*. The column, called "Along the Wine Trail," regularly ran to three thousand words and was often longer. Fougner rarely confined himself to wine; he gossiped about New York restaurants, described special dinners in loving detail, provided long and complicated recipes, and answered readers' questions. He once estimated that he had replied to more than three hundred thousand mail queries during the eight years that he wrote the column.

After his death from a heart attack in April 1941, at the age of fifty-six, *The New York Times* said in an obituary: "No research was too obscure in answering any question from readers interested in the history or validity of any fine point of eating or drinking."

Fougner presented the classic image of a connoisseur of the good life. Portly, balding, and always impeccably dressed, he was known as "The Baron" among friends and in the larger world of food and wine that was his daily beat. During the years that he wrote "Along the Wine Trail," he founded or co-founded no fewer than fourteen wine and food societies, most notably Les Amis d'Escoffier, named after the French chef Auguste Escoffier. He made it a point to preside at the society dinners that were the only reason the clubs existed.

Fougner often said that *The Sun* started the column to instruct the public in the finer points of eating, and especially drinking, that had been lost during the years of Prohibition. He went further, judging culinary contests and advising on the training of waiters and bartenders. At the time of Repeal, he described New York as a gustatory wasteland, but in December 1940, four months before he died, he noted proudly that he needed an entire column just to list what he considered "the greatest gastronomic events of the year."

A typical Fougner column, that of February 21, 1940, begins, like many journalistic endeavors, with a lament about how hard the work is: "In one of the most strenuous weeks on record, so far as wining and dining activities are concerned, three dinners had to be canceled owing to overcrowding of the Trail calendar."

The most regrettable omission, he goes on to say, was a golden wedding celebration for an old friend at the Hotel Kentucky in Louisville. The dinners that prevented the Baron from dashing off to Louisville included "an extraordinarily brilliant" one at the Lotos Club that featured pressed duck and Château Latour 1923, a "rousing sendoff" aboard the Cuba Mail S.S. *Oriente* for a member of the Society of Restaurateurs, a wine-tasting lunch with Albert Fromm featuring Napa Valley wines made by the Christian Brothers, a Champagne tasting with the managing director of the St. Regis, and a "great dinner staged by French Veterans of the World War" at the Hotel Pennsylvania, for which he published the entire menu. (They drank a 1934 Sylvaner with the *sole Marguery* and Château Margaux 1928 with the *poussin Pennsylvania*.)

The rest of the column is an exchange with a reader from Larchmont, New York, who writes: "I have perfected a variation of your 'Lord Botetourt Punch,'" offering it in exchange for "a set of directions for the making of Sherry-wine jelly." Fougner supplies the reader's version of the punch, which involves red wine, brandy, seltzer, and fruit, then goes on to give the instructions for making the jelly.

In spite of his impressive food and wine credentials, Fougner listed himself in *Who's Who* as a newspaperman. He was born in Chicago and came to New York to take a job as a reporter for *The New York Herald* in 1906. He later worked for *The New York Press* and joined *The Sun* in 1912. Shortly thereafter, *The Sun* sent him to London as chief correspondent and European manager. He covered World War I until 1917, when he returned to New York, joined the United States Treasury Department, and handled publicity for Liberty Loans. He rejoined *The Sun* in 1920, left to do public relations and freelance writing, then went back to newspapering once again in 1931.

Accepted wisdom attributes America's current interest in food and wine to tastes our troops acquired overseas in World War II, to easier travel and more leisure time in the postwar years, and to the emergence of a new middle class with money to spend on luxuries. But the Fougner columns indicate that wine affairs, gourmet societies, and a taste for the good life were part of the New York scene in the depths of the Depression.

FEBRUARY, 1992

## A WRITER MANY FRENCH CHÂTEAU OWNERS RELY UPON

A typical wine writer was once described as someone with a typewriter who was looking for his name in print, a free lunch, and a way to write off his wine cellar. It's a dated view. Wine writers now use computers.

It also fails to take into account Michel Bettane. Mr. Bettane is France's preeminent wine writer, and much more. When a winemaker in the Loire is depressed about his most recent vintage, he calls Mr. Bettane. When a chateau owner in Bordeaux is trying to determine when to pick his merlot, or a producer in Beaujolais isn't sure how much sugar he should add to increase the alcohol level of his wine, he will call Mr. Bettane—sometimes in the middle of the night.

Mr. Bettane is an editor of France's best wine magazine, *La Revue du Vin de France,* or *R.V.F.,* as he and everyone involved in the wine business here call it. Since 1996 he and his fellow editor, Thierry Desseauve, have also published an annual critique of all French wines. The 1998 edition, which appeared in September 1997, lists "the 945 best vineyards" and offers comments on 4,164 wines.

The current volume caused a considerable stir in wine circles by failing to award Château Mouton-Rothschild the same three stars, the book's highest accolade, that it bestowed on the other top Bordeaux chateaus: Lafite-Rothschild, Latour, Margaux, and Haut-Brion. Mouton-Rothschild had to settle for two stars.

"Just trying to bring attention to their book," sniffed Baroness Philippine de Rothschild, Mouton's owner.

Mr. Bettane shrugged. "Mouton is one of the glories of French wine," he said. "It will be back in form soon."

Some other chateau owners, whether through jealousy or conviction, were glad to see Mouton take a hit. But several years ago, almost all Bordeaux was outraged when Mr. Bettane joined a group of mostly American and British wine experts in San Francisco to reassess the famous 1855 classification of the top sixty Bordeaux wines.

It was a harmless exercise, bankrolled by the millionaire Gordon Getty. The general conclusion, which surprised no one, was that for all practical purposes, that ranking had withstood the test of time. In France, however, the event was seen as an attack on French culture itself. And Mr. Bettane was seen as an enological traitor. Which he found delightful.

Rather short, rather round, and having an insouciant air, Mr. Bettane could be typecast as a bon vivant. But his appearance belies a sense of dedication rare among his French colleagues and virtually unmatched elsewhere in wine circles.

He has been deeply committed to wine since his boyhood. He was born in Washington, D.C., where his father was a military attaché, and lived in the United States until he was seven. But he remembers holidays in France, poking around in the wine cellar of an uncle's restaurant near Paris. For some reason, the label on a bottle of white Burgundy, a Meursault from the Comtes Lafon, stuck in his memory. Later, in 1973, Comtes Lafon was the first vineyard he chose to visit. He still remembers the wines he tasted there. "The Meursault—Perrières—were sublime," he recalled recently. In 1977 he enrolled in Steven Spurrier's L'Academie du Vin in Paris. Within months he was part of the faculty, teaching young I.B.M. executives the difference between Pouilly-Fuissé and Pouilly-Fumé.

But his real training came at the hands of some of France's best winemakers. Among his favorites: Étienne Guigal in the Rhône Valley, Léonard Humbrecht in Alsace, and Michel Delon at Château Léoville-Las-Cases in Bordeaux. "These men spent entire days with me answering endless questions, most of them stupid," he said. "To really learn wine, you must go to the people who make it. The best of them love to tell you how they work, once they see you are serious." He smiled. "To be taken seriously is a long, long process," he added.

Mr. Bettane is part of a new generation of specialists in a country where until recently wine was an inward-looking, chauvinistic little world. He consults in Italy and visits the United States whenever he can. He admires the wines of Randall Grahm's Bonny Doon Vineyard and the cabernets that Al Brounstein makes at Diamond Creek. Twice he has been to the

Pinot Noir Celebration in McMinnville, Oregon, an annual three-day party for amateur pinot noir lovers. When it ends, the experts—American, French, Australian, Italian—hold several days of private technical discussions.

"I think I'm the only writer ever asked to attend the technical talks," Mr. Bettane said.

Loving America and some of its wines does not prevent him from heaping scorn on one of our obsessions: rating wine by the numbers, usually on the hundred-point scale that we all first experienced in grade school. That system was popularized by Robert M. Parker Jr. in his bimonthly newsletter, *The Wine Advocate,* and later adopted by most American consumer wine publications.

"A wine tastes one way in the morning, a different way in the afternoon," Mr. Bettane said. "I'm constantly surprised by how differently I perceive California merlots and cabernets in America and then in France. The California air, the food one eats there, the way it's seasoned give their wines a fullness that seems excessive in France.

"In the same way, I find some of the best Australian wines far too acidic when I drink them here. Closer to home, I'm always astonished at how badly great Burgundies often seem to perform in Bordeaux, and vice versa."

He continued: "Should we give up judging wine? No. But instead of giving simplistic numbers, we should judge wine in relation to its vintage, to the type of grape from which it's made, and in the case of a very great bottle, by comparing it to its performance down through the years.

"It's wrong to put every wine in competition with every other wine. The Bordelais, who classified their wines on five levels, have a lot to teach us about wine judging."

True enough, but is there a Bordeaux producer who is not thrilled to get a ninety rating, or higher, from Mr. Parker? Call it cultural imperialism, but these days, rating wine by the numbers, American style, is as important in St.-Émilion as it is in Sonoma.

JANUARY, 1998

## NONVINTAGE YEARS

Each day of his three years' captivity in Lebanon, Jean-Paul Kauffmann recited to himself the names of the sixty-one chateaus in the 1855 classification of the wines of Bordeaux. "I'd try to write them down on empty packets from Cedars, the revolting Lebanese cigarettes they gave us," said the French former hostage, "but I lost my list each time we were moved." Mr. Kauffmann's captors moved him eighteen times and eventually took away his pen.

"By the end of 1986," he said, "I began to forget some of the Fourth Growths, invariably Châteaux Pouget and Marquis-de-Terme." As time went on, he began to forget some of the Fifth Growths as well. (The 1855 classification organized the sixty-one chateaus into five groups, by quality and price.)

"Not to know the famous classification by heart was devastating," Mr. Kauffmann recalled. "Was I losing all touch with civilization, becoming some kind of barbarian?"

When Moslem fanatics kidnapped Mr. Kauffmann near the Beirut airport on May 22, 1985, he had just arrived in Lebanon as a correspondent for the French newsmagazine *L'Évènement du Jeudi.* But Jean-Paul Kauffmann was also editor in chief of a small but influential wine magazine, *L'Amateur de Bordeaux,* which is published in Paris four times a year. In the most recent issue, published after his release last May, he recounts his long ordeal in a moving article.

"Could anything be more ridiculous," he wrote, "than the editor of *L'Amateur de Bordeaux* being kidnapped by fanatics of the Koran for whom the very word for wine, *nabid,* is an abomination. . . . I, who had loved the friendly, elegant world of Bordeaux, suddenly found myself in this brutal, elementary world.

"I who had loved the subtle shadows of the *chais,* redolent of the vanilla aroma of new oak barrels, was going to spend three years in a labyrinth of dark cells, a subterranean world of pain and suffering, a universe out of the drawings of Piranesi.

"During the entire three years," Mr. Kauffmann wrote, "I never forgot the taste of wine, even though it was no more than a memory, a fleeting sensation. I was Proust without a madeleine; my madeleine was my memory."

He compared his memory to the armoire of a dedicated wine collector. Each night, while his captors murmured prayers in the next room, he dreamed of Bordeaux. "How often I walked the roads of the Médoc," he wrote. "Fastened to my chain though I was, I never ceased to wander through the lovely hills of the Entre-Deux-Mers. I saw in my mind the rows of plane trees leading to Château Margaux, the Tuscan-style gardens of Yquem, the towers of Cos d'Estournel, the flowers at Giscours."

Through the long hours, he introduced his fellow prisoner, the sociologist Michel Seurat, to the mysteries of wine. "He taught me Kant and Hegel," Mr. Kauffmann said. "I talked of literature and the world of Bordeaux." Later, Mr. Kauffmann was to watch helplessly as his friend wasted away and died of cancer. At one point, the guards admitted another prisoner, the Jewish Lebanese physician Elie Hallak, to the cell to treat Mr. Seurat. Dr. Hallak had read about Jean-Paul Kauffmann and said, with a smile, "After we get out of here, you must teach me about wine." Later, the guards told Mr. Kauffmann that Dr. Hallak had been shot.

Before he became ill, Mr. Seurat received some books: a Bible, the collected novels of Jean-Paul Sartre, and Tolstoy's *War and Peace*. Over and over, the two prisoners discussed a passage in one of the Sartre novels, *Le Sursis*.

Two characters dine at Juan-les-Pins, on the Côte d'Azur, in 1938, drinking Château Margaux. In stirring words, one of the two, a painter, renounces his art and resolves to go off to Spain to join the Loyalists. Chained in their sweltering cell in Beirut, the two prisoners wondered which vintage of Margaux would have been served. Sartre didn't mention one. Eventually they settled on 1934.

"We decided that only a superb vintage could have inspired such rhetoric," Mr. Kauffmann wrote.

Several times, Mr. Kauffmann was moved to another prison in a metal coffin. Once, in 1987, after spending twelve hours in the box, he was sure his jailers had abandoned him. "If I get out of this nightmare alive," he promised God, "I will never again drink alcohol. But then, it occurred to me that life wouldn't be worth much without good Bordeaux, so I compromised on three months."

Even in the depths of his despair, Mr. Kauffmann noted later, he never completely gave up hope. His offer must have been acceptable because he was released from the coffin shortly thereafter. (True to his word, for the first three months after his release, he drank no alcohol.)

Two other hostages, Marcel Fontaine and Marcel Carton, both officers at the French embassy in Beirut, turned up later in 1985, to join what Mr. Kauffmann called "our fraternity of misery." Mr. Kauffmann undertook to teach his two new pupils about wine. Mr. Seurat was soon too ill to join in. "I don't feel like talking," he told the others, "but I enjoy listening to you."

Mr. Fontaine enjoyed Champagne, so Mr. Kauffmann composed a crossword puzzle for him that included the word "Ay." Ay is a famous Champagne village, the home of Deutz & Geldermann and Bollinger; it is also, for obvious reasons, a popular word in French crossword puzzles.

Mr. Carton knew and loved the town of Langon, not far from Sauternes, at the southern end of the Bordeaux region, and had occasionally bought wine from Pierre Coste, a prominent Langon wine shipper. Mr. Carton liked to recall one of his favorites: Château L'Étoile, an attractive white Graves.

"We knew hunger, cold, heat, fear," Mr. Kauffmann wrote, "but we never stopped talking about wine. It was our last connection with the world of the living."

NOVEMBER, 1988

## A REPORTER'S REPORTER

Pierre-Marie Doutrélant died in 1987 while jogging in the Bois de Boulogne, in Paris. He was forty-six years old, left a wife and a couple of kids, and may have been the best wine journalist of our time.

Most of what you read on the subject is the work of wine writers. There are some good ones, but they are, for the most part, committed. One expects them to tell us nice things, and they rarely disappoint. Philosophically, wine writers are much alike; they vary only in degrees of felicitousness.

The reporter's job is different, or should be. The reporter's tools are an eye for detail and a supply of skepticism. He might love wine—no harm in that—but he should love a good story more. Here is Doutrélant quoting a Beaujolais public relations man:

"'We don't bribe the papers, but we do have good journalist friends who never miss an opportunity to come see us when they're in the neighborhood.'"

Or a government inspector on the illegal use of sugar to boost the alcoholic content of Beaujolais-Villages: "'If the law had been enforced in 1973 and 1974, at least a thousand producers would have been put out of business.'"

These unchauvinistic remarks were published in 1975, a time when hard reporting on wine was almost nonexistent in France. Food and wine scribes in La Belle France routinely cross the line, working as much for the industry as they do for the press. This kind of talk was sacrilegious.

Almost thirty years ago, Doutrélant disclosed that many famous Champagne houses, when short of stock, bought bottled but unlabeled wine from cooperatives or one of the big private-label producers in the region, then sold it as their own. He explained how the growers of Côtes du Rhône planted mourvèdre and syrah, two low-yield grapes that give the wine finesse, strictly for the benefit of government inspectors. Then, when the inspectors left, they grafted cheap high-yield vines—grenache and carignan—back onto the vines.

Pierre-Marie Doutrélant came from Hazebrouck, in the north, near Lille. He worked on a paper in Angers, then moved to Paris and *Le Monde,* where for eight years he covered politics, urban affairs, and, now and then, wine. Some of his wine articles, including those mentioned above, appeared in his first book, *Les Bons Vins et Les Autres* (*The Good Wines and the Others*). His section on Chablis is subtitled "Or how the public authorities decided that the best way to combat fraud was to make its practice legal."

He did politics for the weekly *Le Nouvel Observateur,* then moved to the weekly *L'Express,* where he took over the section called "Portraits," a series of profiles he wrote on people in the news. According to another Paris journalist, Doutrélant was slated shortly to take over *L'Express Paris,* a sort of grown-up *New York* magazine for the City of Light.

While he covered economics, politics, lifestyles, and everything else from week to week, his not-so-secret passion was food and wine. His acerbic comments on the pompous three-star chefs appeared regularly in *Cuisine & Vins de France* and last year were compiled in his second book, *La Bonne Cuisine et les Autres.* Among the immortals he skewered like a *brochette d'agneau* was the great Paul Bocuse himself.

French reporters tell the story of Bocuse about to meet a journalist of whom he had never heard. "Is he," asked the master nervously, *"du genre Doutrélant?"* ("a Doutrélant type?").

One of Doutrélant's best pieces was a profile of the publicity-hungry Bernard Loiseau, a two-star chef and restaurateur at La Côte d'Or in Saulieu, on the northern edge of Burgundy. Not a few observers of the cutthroat world of big-time French cooking credit—or blame—the Doutrélant profile for delaying Mr. Loiseau's third Michelin star.

Mr. Loiseau, the spiritual heir of the great Alexandre Dumaine, who ran

La Côte d'Or before World War II (and taught Bocuse), had come up with something he called *"la cuisine du vapeur"*—steam cooking. The result, according to Pierre-Marie Doutrélant, was sauces that looked and tasted like water. He suggested that the money Mr. Loiseau invested in a heliport to lure Parisian fat cats—and impress the Michelin inspectors—might better have been spent on more substantial food.

Doutrélant was of the school of French reporters who came up in the 1960s, who were on the barricades in 1968, emotionally if not literally. They saw through the ossified political establishment the students hoped to bring down and through the hypocrisy of the students themselves, who rioted on weekends and after exams. I don't think I ever saw him wear a tie—or for that matter, a suit—even when, briefly, he wandered through the halls of this newspaper en route to the vineyards of California. Typically, the Napa Valley scene, half work and half chichi, amused more than it impressed him.

We never worked a wine story together. I saw him from time to time at political events, especially during the French presidential campaign in 1980. When the Socialists won, their headquarters in the upper-crust Rue Solferino became one big victory party. Pushing through the crowd, I suddenly came face to face with Pierre-Marie Doutrélant, his hair more disheveled than ever, his face crinkled into a huge smile that virtually closed his eyes.

There was no doubt where his sympathies lay. But his writing on those events was cool and detached. It was the same with wine. He even had some vines at his weekend place in the hauntingly lovely Gers. But that's as far as it went. When he wrote, he was tough and uncompromising, be it on the Médoc or Mitterrand. For a reporter, that's the way it's supposed to be.

NOVEMBER, 1987

## "I GOT TO DRINK MORE GOOD WINE THAN MOST . . ."

In the autumn of 1927, A. J. Liebling spent a week in and around Dijon and Gevrey-Chambertin, eating well and drinking wine. Thirty-five years later he recalled the time as his "true initiation into the drinking of Burgundy."

The late Abbott Joseph Liebling's reputation as a gourmand and connoisseur rests unassailable. When he died they retired the knife and fork. I had read most of his food writing, but none of it had much to do with wine. I had long been curious about what he drank and whether he wrote about wine with the same flair he brought to food.

Then one day, browsing through remaindered books, I came across an anthology that pulls together four collections of Liebling articles, including *Between Meals,* the book into which he put just about everything he wrote on wine, perhaps half a dozen articles. Not much, but like everything he did, they were pretty good.

Some writers, like some wines, age well, and the serious wine buff should be familiar with the best of both. Herewith, then, a sampling—a *dégustation,* if you will—of Liebling on wine.

He says he started learning about wine during his year as a Sorbonne student in 1926 and 1927. I don't believe him. He had been traveling to Europe on and off since childhood. His father, a well-to-do furrier, was no connoisseur but enjoyed the good life. The son had to have known something about the grape before he arrived in France in 1926.

On the other hand the beginning, as he presents it, was not auspicious. As a "student" (he quit classes after three weeks), Liebling trudged across Paris each week from the Rue de l'École de Médecine, near the Odéon, where he lived, to the vast headquarters of the Crédit Lyonnais in the Boulevard des Italiens, where he picked up the allowance money sent by his father.

Along the way he often stopped for lunch at the Restaurant des Beaux Arts, in the Rue Bonaparte, where, of all things, he drank Tavel rosé bottled on the premises. A sign outside—I think I've seen it there—read "Renowned Cellar—Great Speciality of Wines of the Rhône." Strictly speaking Tavel *is* a Rhône wine, but it is hardly the right beginning for a serious wine-drinking career.

Liebling's recollection of the wine thirty-five years later was mellow. "Tavel," he wrote, "has a rose-cerise 'robe' like a number of well-known racing silks, but its taste is not thin or acidulous, as that of most of its mimics is. The taste is warm but dry, like an enthusiasm held under restraint, and there is a tantalizing suspicion of bitterness when the wine hits the top of the palate. With the second glass, the enthusiasm gains; with the third it is overpowering. The effect is generous and calorific, stimulative of cerebration and the social instincts."

Lovely prose for an indifferent wine. One would have to agree, though, that the price was right: 12 cents for a half-bottle of Tavel; 14 cents for the Tavel *supérieur.*

On those days when he was fairly sure there would be money at the Crédit Lyonnais, Liebling drank better—thank Heaven. True to its enameled plaque, the Beaux Arts offered a full line of Rhônes up to and including both white and red Hermitage, which, if his memory was accurate, sold in 1926 for about 20 cents more a half-bottle than the lowly Tavel.

His favorite, though, was Côte Rôtie; it was, he wrote, "my darling." "Drinking it," he went on, "I fancied I could see that literally roasting but miraculously green hillside, popping with goodness like the skin of a roasting duck, while little wine-colored devils chased little nymphs along its simmering rivulets of wine."

Three decades later Liebling had a "return match with Côte Rôtie" at Prunier's in London. "I approached it with foreboding," he wrote, "as you return to a favorite author you haven't read for a long time, hoping he will be as good as you remember. But I need have had no fear. Like Dickens, Côte Rôtie meets the test. It is no Rudyard Kipling in a bottle, making one suspect a defective memory or a defective cork."

By the end of what he cavalierly referred to as the academic year, Liebling was drinking much better. He decided to return home to his job at *The Providence Journal* by ship from Marseilles. The route there offered enticing possibilities to the young gourmand.

"I descended from the Paris–Marseilles express at Dijon," he related, "to eat the first rear-guard action of my anabasis. At the Restaurant Racouchot, in the Place d'Armes, I made the acquaintance of *caille vendageuse* (quail with grapes) and drank a bottle of Corton–Clos du Roi."

On the waiter's recommendation he moved into a small auberge nearby in Gevrey-Chambertin, run by a retired army officer named Robaine. For a week Robaine escorted his young guest around to the great cellars of Chambertin, Fixin, and Vougeot, explaining to the winemakers that Liebling was a rich bootlegger come to buy wine for a New York nightclub.

"I got to drink more good wine than most men are able to pay for in their lives," he recalled, "and Robaine drank along with me, pushing the merchandise as he drank, and winking grossly at the proprietors of the vineyards to indicate that he was conspiring with them to get a good price from me."

On one afternoon Liebling worked his own little scam in a restaurant at Nuits–St.-Georges, a few miles down the road from Gevrey. With his lunch he drank a superb bottle of Grands Échezeaux. When the young man across from him, a manager for a local wine merchant, asked his opinion of the wine, he allowed that it was excellent but not extraordinary and that he had drunk as well in Paris and even Ireland.

Stung, the young man said: "I cannot tolerate that you should carry away such a mediocre impression of our cellars. I invite you to sample what we call good wines at our place." And he added: "Between you and me, the fellow who bottled that, although he is my boss's cousin, is a sharp chap. Doubtful integrity."

"The afternoon I spent in the cellars of his firm was one of the happiest of my life," Liebling wrote years later. "I regret that I have forgotten the firm's name. I was lucky to remember my own."

Bottle followed bottle. Each, he conceded, was a bit better than the last. The last bottle, a Romanée-Conti, was, he recalled, first rate. "Well worth the voyage from North America to taste," he said. "Thunderously superior to that stuff I had with lunch. My benefactor was pale with gratitude. But the bottle at the hotel had been the best of the day."

There is more. But this is a good sampling. They don't make wine like that anymore. They don't write like that either.

APRIL, 1984

# CHAPTER 2

# TENDING THE VINE AND MAKING THE WINE

n the first years of America's wine boom, back in the late 1950s and 1960s, new California winemakers were guided by the weather. This is a good spot for a vineyard, someone would decide, because it's hot during the day and cool at night—perfect for grapes. The soil didn't matter because all California soil was good.

It took the Europeans thirty years or so to convince the Americans that the soil was in fact everything. Once the winemakers were the stars. Now every American winemaker says that wine is made in the vineyard. Now, as a story about a tasting put on by Francis Ford Coppola shows, the shoe is on the other wine-stained foot. A group of young Bordeaux winemakers is closely imitating some of the best small wineries in the Napa Valley, with outstanding results.

Some things don't change, as typified by the picking at Château Margaux and the annual ritual of the *assemblage* at Château Prieuré-Lichine. Some traditions endure but these past thirty years have been a time of intense experimentation and innovation everywhere in the wine world. Nicolas Joly, experimenting with biodynamism in the Loire; the Rhone Rangers, getting away from cabernet and replicating the wines of the Rhone Valley; Dick Graff, off in the remote Gavilan

Mountains, re-creating true Burgundian-style pinot noirs; Paul Draper, in Santa Cruz, making a world-class wine from the lowly zinfandel grape, and, bringing two worlds together, there are the Mondavis and the Rothschilds, together producing the exceptional wine known as Opus One.

And, finally, symbolizing the astonishing globalization of the wine business in recent years, there is Michel Rolland, the most prominent of a new breed known as flying winemakers.

## BORDEAUX EDGES CLOSER TO THE NAPA VALLEY

Francis Ford Coppola marked his twenty-fifth anniversary in the wine business in September 2000 with a daylong party at his Niebaum-Coppola Winery in the Napa Valley. The festivities included a dinner with vintages of his prestigious Rubicon red wine dating back to 1978, served in magnums, followed by a screening of his film *One from the Heart.*

For me and a group of other guests, the highlight of the day was a tasting comparing California and Bordeaux cabernet sauvignon blends from the 1985 and 1995 vintages. Led by Niebaum-Coppola's winemaker, Scott McLeod, we tasted 1985 Bordeaux from châteaus Lynch-Bages, Pichon-Lalande, Léoville–Las Cases, Haut-Brion, Margaux, and Lafite-Rothschild. From the Napa Valley, the '85s were Heitz Cellars Martha's Vineyard, Niebaum-Coppola Estate Winery's Rubicon, Joseph Phelps's Insignia, Stag's Leap Wine Cellars Cask 23, Château Montelena Cabernet Sauvignon, and Robert Mondavi Reserve Cabernet Sauvignon.

The 1995 wines were the same except for the Heitz. Like most Napa Valley wineries, the Heitz Cellars vineyards were seriously damaged by phylloxera in the late 1980s and replanted in 1992. Another Heitz estate wine, Trailside Vineyard, took its place.

The first flight of wines in the tasting on Saturday morning consisted of the dozen '85s, all identified, with the six California wines in one row and the Bordeaux wines in another. In the second flight, the wines were served blind.

To me, the remarkable feature of the tasting was the change in style among the French wines. The 1985s offered no surprises. There were some excellent wines, all true to form. They were elegant and restrained; the California wines were full-bodied and lush. The French were more angular and austere; the Californians more rounded and mellow.

Ten years later, all that had changed. The French wines had become, well, more Californian. Mr. McLeod asked the group—winemakers, a couple of restaurant people, a few journalists, and one prominent filmmaker—to guess which 1995s were Californian and which were French. Only one person got them all correct. I got five wrong out of twelve, but then I've always considered humiliation an important component of wine judging.

My favorites among the 1985s were the Lynch-Bages, the Margaux, and the Cask 23. Among the 1995s I liked the Château Montelena, the Phelps Insignia, and the Léoville–Las Cases.

At the dinner, I especially enjoyed the 1978, 1981, and 1984 Rubicons.

Both 1985 and 1995 were fine vintages in Bordeaux and California. The 1985 Bordeaux lacked something of the power and concentration of 1989 and 1990. And yet they seemed, at this tasting anyway, harder and less generous than the Napa wines. In California, 1985 is thought to have been the best year of the decade. Both the Bordeaux and the California wines matured fairly early. As for the 1995s, most critics rated the Bordeaux just below the more tannic 1996s, and the California wines just below the 1994s but above the 1996s, all three being excellent years.

Bordeaux and the Napa Valley, and particularly Niebaum-Coppola, are inextricably intertwined. Bordeaux has always been the inspiration for Napa cabernets, and both regions use the same grapes. Gustave Niebaum, who founded the estate in the late 1800s, used Bordeaux cuttings to start his vineyard and, as Mr. Coppola noted before the tasting, they "still plant the Niebaum clone of cabernet sauvignon today." In fact, Niebaum-Coppola's Rubicon was one of the first ultrapremium California wines to be made from a blend of Bordeaux varieties.

An earlier generation of Napa Valley vintners boasted that their best wines were 100 percent cabernet. Now most winemakers agree that the best wines in California, as in Bordeaux, are blends. Among the 1985s, the Rubicon is 61 percent cabernet sauvignon, 26 percent cabernet franc, and 13 percent merlot; the Phelps Insignia is 60 percent cabernet sauvignon, 15 percent cabernet franc, and 25 percent merlot. There are, of course, traditionalists. The Heitz Martha's Vineyard 1985, like all previous Martha's Vineyards, is 100 percent cabernet sauvignon.

The Bordeaux winemakers have always blended and have never bothered to explain their blends. Among the wines in this tasting, Lafite-Rothschild usually has 70 percent cabernet sauvignon, 25 percent merlot, and small amounts of cabernet franc and petit verdot. Lynch-Bages and Margaux are similar.

After the tasting someone observed that when experts identify Bordeaux wines as Californian and Californian wines as Bordeaux, the days of comparing one region with another may be ending; that the convergence of styles, so evident in the tasting, may mark the beginning of a time when we will compare California estates with Bordeaux chateaus—with no regional bias.

"The world is a smaller place," Mr. McLeod said, leaving no doubt that he feels California will occupy the premier position. "We are entering a golden age here. The first decade of the new millennium may well belong to the Napa Valley."

SEPTEMBER, 2000

## SEEDS OF A VINEYARD REVOLUTION

Nicolas Joly had put in two years with Morgan Guaranty in New York, advising on investments in chemical companies, then a year in London. In 1976 the bank wanted to move him once more. He quit instead, and returned home to another bank, that of the Loire River, and to his family's estate at Coulée de Serrant, a tiny hamlet fifteen minutes by car from Angers, the regional capital.

"I was sick of numbers," said Mr. Joly, whose M.B.A. diploma from Columbia is now, he thinks, stuffed in a drawer somewhere with old seed catalogues. "It took all my diplomatic skills to explain to my wife, who's the daughter of a German diplomat, that she'd gone to bed with a banker and awakened with a farmer."

Mr. Joly's late father, André, a physician in Angers, bought the rundown estate, Le Clos de la Coulée de Serrant, in 1962 as a weekend place. He had no idea that he had bought a famous Savennières vineyard or that Curnonsky, the renowned French gastronome, once ranked the wine among the best five white wines he'd ever tasted.

"Once we realized what we had, we began to improve it," said Denise

Joly, Nicolas's mother, who calls herself a winemaker by chance. "It's been a long and difficult process, and we've loved every minute of it."

The process has only begun. In a bookstore in 1981, Mr. Joly, then in his mid-thirties, found a work of the Austrian philosopher Rudolf Steiner, the father of the theory of biodynamics. "I took the book on a skiing trip," Mr. Joly said, "and it changed my life."

Biodynamics posits an agriculture based on harmony and respect for nature. "It's something like homeopathy," Mr. Joly said, "except that you have to make your own medicine."

Two decades ago, the seventeen-acre Coulée estate was overrun with rabbits. Applying Steiner's principles, Mr. Joly burned a fresh rabbit skin down to a handful of ashes. The ashes were mixed with water and sprayed on the vineyards. There have been, he swears, no rabbits since.

Many of Steiner's ideas are consistent with the organic farming practiced by people who have never heard of the Austrian, who died in 1925, or his biodynamics. Mr. Joly uses only natural fertilizer and some copper sulfate, the famous Bordeaux mixture, against mildew.

"It takes seven years to bring back a vineyard, to reverse the effects of years of artificial fertilizers and pesticides," Mr. Joly said, "so we are just beginning to see the benefits of our work here. But what a difference!"

Steiner says all good agricultural practices are governed by the planets and even the time of day. "Suppose we are working under Sagittarius, the sign of warmth," he said, "but in two days we will move under another sign, one less propitious for our work. We work hard, then take time off from the vineyards."

Mr. Joly said his first vineyard manager quit. "He had been trained with pesticides, weed-killers, and all the other chemicals," he said. "He couldn't accept what he saw as a step backward. Ah, but the workers, the older fellows, the real countrymen—they understood immediately. They had heard their grandmothers say that wheat should be sown only under a waning moon."

The wines of Coulée de Serrant are made from the chenin blanc grape. They bear no resemblance to California chenin blanc, which, with few exceptions, is semisweet and undistinguished. The main problem with a good Savennières, and particularly with those of the Coulée de Serrant and its neighbor, La Roche aux Moines, is longevity. The young wines are highly acidic, with little bouquet until they are three years old, and are really not worth drinking until they are ten years old. No one has determined how long they will last, but wines from the 1940s and 1950s, when the vineyards were shoddily maintained, are still fresh-tasting. The oldest surviving Savennières are sweet wines in the style of Sauternes.

The buzzword of modern viticulture is *clone*. Few serious modern vineyard keepers will admit to not having a special grape clone or at least being involved in clonal selection—all for quality, quantity, and healthier plants.

Nicolas Joly is dead set against clonal selection. He believes that no matter how good a single vine is, if it is reproduced a thousand times it cannot compare with a family of vines, each contributing something of its special character to the whole, all together slowly adapting to the vineyard site.

Gradually, Mr. Joly and his small team of workers have replanted old vines and cleared woods that were vineyards in the 12th century. It is backbreaking work. Most of the vineyards sloping to the Loire are so steep they must be worked with horses. Slowly, production has increased. About three thousand cases a year now find their way to three-star restaurants in France; most of the wine is exported.

On a rainy March day, the Loire is shrouded in mist. The ancient land is a study in gray and ocher, punctuated by the unnaturally bright reds and yellows of primroses. Halfway down the slope a plowman and his horse struggle through the shale and earth.

Aside from the railroad tracks along the river, little has changed since monks tended these vineyards six hundred years ago. Mr. Joly says those monks enjoyed a harmonious relationship with nature that no longer exists, and that he hopes to restore.

In that respect, one great challenge remains: to restore Coulée de Serrant so that the ungrafted vines those monks cultivated can thrive again. Vines grew naturally for thousands of years in Europe, until the invasion of the phylloxera louse from the United States in the late 19th century devastated all the vineyards of Europe.

The only known antidote is phylloxera-resistant American vine roots. All European vines are grafted onto American roots. Purists still maintain that current vintages cannot match pre-phylloxera wines. Mr. Joly is betting that by revitalizing his soil, by restoring the microbes and minerals that give it its life, he will produce vines with the vigor to resist phylloxera.

"The soil is dead," he said. "For many feet down it is lifeless. The vine reaches down to find nourishment and none is there. So it depends on synthetics and is prey to diseases that require even more synthetics."

His critics say his techniques would be unworkable on any larger, more commercial vineyard. His labor costs, they say, are double those of most of his neighbors, and his yields are often half theirs. And his enthusiasm for astrology raises more than a few eyebrows. But growers and winemakers from all over Europe come to Savennières to learn his techniques. No less a wine star than Lalou Bize-Leroy, of the Domaine Leroy in Burgundy, has begun to employ many Joly techniques.

"Oh, yes," Mr. Joly said, "some think I am a madman. But even then, they admit that so much of what I say makes sense. I prefer to let the wines speak for me."

APRIL, 1991

## IDENTIFIED FLYING OBJECT

Are you bored with your usual wines? The same old Lafite-Rothschild, Échezeaux, and Dom Perignon, week in, week out? Perhaps it's time for a little flying cigar.

Yes, flying cigar. It's a wine—but why hear it secondhand? Let Randall Grahm tell it. He's the chap who made the stuff, and what follows are the notes from the back label, which he wrote: "In 1954, the village council of Châteauneuf-du-Pape was quite perturbed and apprehensive that flying saucers or 'flying cigars' might do damage to their vineyards, were they to land therein. So, right-thinking men all, they passed an ordinance prohibiting the landing of flying saucers or flying cigars in their vineyards. (This ordinance has worked well in discouraging such landings.) The ordinance further states that any flying saucers or flying cigars that did land were to be taken immediately to the pound."

The French for flying saucer is *cigare volant,* and that's what Mr. Grahm calls his wine.

A bit fey? Perhaps. But not entirely irrelevant. Châteauneuf-du-Pape is probably the best-known wine of France's Rhône Valley, and Mr. Grahm has sought to emulate it in his Cigare Volant. He is the owner and winemaker at Bonny Doon Vineyard in the Santa Cruz Mountains, south of San Francisco.

Mr. Grahm, like many California winegrowers before him, discovered early on that the California wine country isn't really like Bordeaux. On a lot of summer days, even in the so-called cooler northern counties, the temperature is more reminiscent of El Alamein than St.-Émilion.

What much of the California vineyard country resembles is the southern Rhône Valley. The Rhône River rises in the Alps and wanders more or less aimlessly through Switzerland and eastern France until it gets its act together east of Lyons, joins up with the Saône River, and heads south 190 miles to the Mediterranean Sea.

The northern Rhône region, from Lyons to Valence, is temperate, but from Montélimar to Avignon the country is meridional—hot and dusty, with red tile roofs everywhere. This is the country of Châteauneuf-du-Pape, of Côtes du Rhône and Gigondas, a land where the syrah grape is king and where a lot of Californian winemakers could feel at home. Particularly those who make petite sirah. Before Randall Grahm, dark, heavy petite sirah was California's most Rhône-like wine. Actually, the syrah and the petite sirah are not the same grape. The California petite sirah grape is known in the Rhône Valley as the *durif.*

Cigare Volant is made from 72 percent grenache, 25 percent syrah, and 3 percent mourvèdre, the traditional mélange used in Châteauneuf-du-Pape.

Grenache, which shows up all around the Mediterranean basin, has been used in California for years, mostly in rosé wines. Mourvèdre, originally a Spanish grape, is grown in a few vineyards around Concord in Contra Costa County in California.

Mr. Grahm first went to Santa Cruz to make pinot noir, the wine of Burgundy. His forty-eight-acre property at Bonny Doon turned out to be too hot. Which didn't stop him from making pinot noir. He bought the grapes each year in Oregon.

He makes Bordeaux-style claret and Burgundy-style chardonnay, but more and more, he has moved to the Rhône-style wines. His young vineyards—planted in 1980—contain plots of marsanne, roussanne, and viognier, the grape of France's most exotic little wine region, Château-Grillet. Marsanne is the grape of Hermitage white, and Mr. Grahm predicts that one day marsanne will be among the principal white wines of California.

One of his favorite grapes is pinot blanc. Actually, it isn't pinot blanc at all, but *melon,* a popular grape in the Loire Valley.

While his own vines are maturing, Mr. Grahm purchases most of his grapes. His syrah, for example, comes from French vines grown around Paso Robles, in Central California.

The Bonny Doon vineyards are as unusual as the wines they produce. As in some regions of France, the vines are planted close together and are not irrigated. "The vines compete with each other," Mr. Grahm says. "It forces them to produce their best. Typically, the best vineyards in Napa produce about six tons of grapes per acre; the best vineyards in Châteauneuf produce two. Is it any wonder that so many of our grapes, and therefore our wines, lack acid and flavor?"

Bonny Doon produced about eight thousand cases of wine in 1985, and may go to around 10,000 this year. Not very much. And since Mr. Grahm makes so many different kinds of wine, there are very few cases of any one type. The current list includes Cigare Volant; syrah; a red blend called Claret; a muscat canelli, a dessert wine made from grapes left out to dry before fermentation; a cabernet sauvignon made from grapes grown in Mendocino's Anderson Valley; a pinot blanc; a *vin gris* made from mourvèdre; and a controversial high-acid chardonnay that last year sold out at the winery.

Mr. Grahm is a city person (Los Angeles) who gave up philosophy for wine. "Actually, I was supposed to become a doctor," he says. "Instead I became an alchemist, which is what a winemaker really is." He didn't give up philosophy completely. One of his 1985 efforts is a white table wine called Le Sophiste–Cuvée Philosophique. It's a blend of marsanne and roussanne. "I really haven't thought out the name yet," says Mr. Grahm, the former philosophy student, "except that a lot of thought went into making the wine."

JANUARY, 1987

## A VINEYARD THAT ATTRACTED A SPECIAL BREED OF WINEMAKER

The death of Richard Graff in a plane crash in January 1998 closed another chapter in the remarkable history of Chalone Vineyards, one of California's better-known sources of fine wines.

Mr. Graff, who was sixty when his Cessna 182 crashed near Salinas, California, apparently because of a mechanical failure, was often described as the founder of Chalone, a 160-acre vineyard in the remote Gavilan Mountains 150 miles south of San Francisco. It was Mr. Graff who brought Chalone into prominence in the late 1960s with his exceptional Burgundy-style wines, but he was far from the first person to try to exploit the vineyard's genuine but elusive qualities.

A vineyard unlike most others, Chalone attracted winemakers unlike most others. Mr. Graff studied music at Harvard University, spent three years as a Navy officer, and was working in a bank when he first saw the place. When he eventually took it over, he succeeded Rodney Strong, who before becoming a winemaker had been a professional dancer, appearing in Broadway shows and choreographing nightclub acts in Paris.

Mr. Strong, who went on to create Rodney Strong Vineyards in Sonoma County, one of California's better-known premium wineries, had succeeded Philip Togni, an Englishman who started his professional life as a petroleum engineer in Peru. He then made wine in Algeria and Bordeaux before coming to California and the Napa Valley in the late 1950s. Mr. Togni, too, eventually created his own winery, Philip Togni Vineyard, on Spring Mountain, also in the Napa Valley.

Chalone's story began before World War I. Old-timers in Monterey County say the site was developed by a Frenchman who had been searching California for soil that resembled the limestone of Burgundy. Having found the place—no one knows how—he still did not plant Burgundy grapes. No one knows what he planted.

At some point, a man named William Silvear bought the property. He leased it out during Prohibition, then revived the vineyard after Repeal. Leon D. Adams, a wine historian who died in 1995, once said that Mr. Silvear's pinot noir grapes were so good that Almaden Vineyards and Wente Brothers both paid premium prices for them in the 1940s and early '50s.

Then the property seems to have fallen into disuse for some time. In the late '50s it was purchased at auction by several San Francisco wine lovers. In 1960 they hired Mr. Togni as winemaker; he had recently been dismissed by Mayacamas Vineyards in the Napa Valley. Mr. Togni had a degree in enology, which he had earned in Bordeaux. "We made an old chicken coop into a little winery, but we sold most of the grapes," Mr. Togni said recently. "So far as I can recall, no wine was made up there until 1966. It was brutal work. Water had to be trucked in, and because the electric lines ended about halfway up the mountain, we used an old generator to run what little equipment we had."

Mr. Togni stuck it out for three years before making a big leap to E. & J. Gallo, then as now the largest winery in the world. "I was getting married," he said. "My wife-to-be took one look at Chalone, and I left soon after that."

Mr. Strong had seen his dancing career winding down and decided to get into winemaking, a calling his ancestors once pursued in Germany. At first he bottled bulk wine and sold it to tourists in Marin County, California. Then with money supplied by some wealthy dancing students, he took over an old winery in northern Sonoma County called Windsor Vineyards. Some of those investors also had an interest in Chalone. When Mr. Togni departed for Gallo, they asked Mr. Strong to take over.

"I leased Chalone," Mr. Strong said, "and for three years, my wife and I drove down there every weekend to work the vineyard, four and a half hours each way. I'd take some of my students—I was also teaching winemaking at Sonoma State College—and any hippies we could find along the way."

Mr. Strong trucked the Chalone grapes north to Windsor, where they were turned into wine and bottled under the Chalone name. One of Mr. Strong's backers was Richard Graff's father, Russ, who knew his son was not happy working in a bank. "His father asked me if I could give him weekend work at Chalone," Mr. Strong said. "He came up and fell in love with the place."

Mr. Graff spent a year studying enology and in 1965, with a loan from his mother, bought Chalone. He made his first wine a year later, but it wasn't until 1969 that Chalone was released commercially.

Chalone's chardonnay and pinot noir developed a cult following. Using that popularity, Mr. Graff and his partner, W. Philip Woodward, a former accountant, attracted new investors, many of them more interested in buying the scarce wine than in a quick return on their money.

In the 1980s Chalone acquired two other wineries: Acacia, in Napa County, and Carmenet, in the Sonoma Valley. Chalone now owns 50 percent of Edna Valley Vineyards, in San Luis Obispo County, a joint venture with Paragon Vineyard Company; 51 percent of Canoe Ridge Vineyard in Washington State; and 24 percent of Château Duhart-Milon in the Bordeaux region of France.

Chalone went public in 1984. Since then, the company has had good years and some less good ones. But even in poorer years, the annual stockholders' meeting, held at the winery, was a happy event, with each announcement of a financial setback greeted by laughter and applause. And why not? Good Chalone wine was served, as was Château Lafite-Rothschild. (Domaines Barons Rothschild, the owner of Lafite and a Chalone share-holder, offers Chalone owners special prices on Lafite and other Rothschild wines and on trips to wine regions.)

Still, there are shareholders who like to make money, and even before Richard Graff's death there were signs of change. He had been out of the day-to-day management for several years and had been making a new wine under his own name. What's more, it had been announced that Tom Self-ridge, a prominent wine industry executive, would move from successful Kendall-Jackson to Chalone as president later that year.

Chalone is getting serious. But without its resident dreamer, it may not be as much fun.

JANUARY, 1998

## SEIZING THE MOMENT TO REACH FOR THE VINE

David Finlayson stood in the doorway of the immense vat room at Château Margaux one day in mid-September 1995, looking out at the rain that would fall intermittently over most of the Médoc, the wine area north of the city of Bordeaux, during the two weeks or so it would take to harvest the grapes.

Not a lot of rain to be sure, but enough to turn what might have been a spectacular vintage into one the experts were calling, guardedly, very good but not outstanding.

Mr. Finlayson, then twenty-three, was getting his first taste of Bordeaux's notoriously uncooperative fall weather. The son and nephew of South African winemakers, he was spending a privileged month at Château Margaux, one of the region's most famous properties, on an apprenticeship that had begun in August.

"I've learned more in a few weeks here than I had in the previous ten years," Mr. Finlayson said, forgetting that ten years earlier he was only thir-teen.

While Mr. Finlayson was asked to work in the cellar, some two hundred people in Bordeaux, mostly students, got letters from Château Margaux on

September 15, inviting them to help in the harvest. They would pick grapes under the hot sun—or in the rain—for about $8 an hour, plus breakfast, lunch, and transportation to and from Bordeaux. For his work, Mr. Finlayson got only room and board.

"If you want to work," the letter said in effect, "show up tomorrow." Few failed to.

"It says something about the economy," said Paul Pontallier, the manager of the chateau. "Or perhaps something about Margaux."

Margaux is one of the five first growths—the others are châteaus Haut-Brion, Lafite-Rothschild, Latour, and Mouton-Rothschild—that sit at the very pinnacle of prestige among Bordeaux's more than three thousand wine producers. Margaux is the only famous chateau to bear the name of the town where it is situated.

Grape picking in the Bordeaux region has become a sought-after job. "Ten years ago, the chateaus were desperate for pickers," Mr. Pontallier said. Now there is a waiting list of people who want to pick, perhaps because there are fewer jobs; more than 50 percent of all the grapes in Bordeaux are now brought in by mechanical harvester.

Picking at Château Margaux began on September 16, a Saturday. Spirits were high. It had been a spectacular summer for grapes in Bordeaux, hot and dry. If the rain held off, it could be a very fine year, even spectacular. Bordeaux was ready for a great vintage; its last great was in 1990.

The decision to pick on the 16th was made on the night of the 13th, and the letters were dispatched on the 14th. But the decision to begin was not lightly made.

"Deciding on the harvest date is a long process," Mr. Pontallier said. "We start at the time of the flowering of the vines, which in 1995 was on June 1. On average, the harvest starts 110 days later. Two weeks before the 110 days are up, we begin sampling the berries. We test for sugar and acid content, but mostly we taste. And when we think the time is ripe, we go."

In fact, the harvest at Château Margaux, or a small part of it, begins earlier. Alone among the first-growth chateaus, Margaux makes a rare and expensive white wine called Pavillon Blanc. Its sauvignon blanc grapes mature before the merlot, cabernet sauvignon, cabernet franc, and petit verdot that go into the chateau's even more famous red.

Pavillon Blanc grapes are harvested by a team of Margaux employees and people from villages in the Margaux area, many of whom have picked for the chateau for generations. This team also works the red grape harvest, bringing the number of pickers in 1995 to about 170.

After working through Saturday the 16th, the pickers were told to take Sunday off and be back on Monday. Many of the merlot grapes needed to mature a bit more. The pickers, divided into four teams, worked Monday and Tuesday. Then the rains came. No grapes were picked on Wednesday or Thursday, the 20th and 21st, but by then all the merlot had been harvested, most of it in very fine condition.

On Friday and Saturday, the teams were out again, moving into the cabernet sauvignon and cabernet franc plots. After a Sunday off, they worked three days before the rain returned. Friday the 29th dawned warm and sunny, and the good weather held through October 2, when the few remaining plots were picked.

The pickers then left, but the fermentation vats were full and the serious work of making one of the world's most-sought-after wines was beginning.

By law, Margaux may make as much as 5,700 liters of wine, or about 1,500 gallons, from every hectare of grapes. (A hectare is about two and a half acres.) The chateau makes 4,800 liters per hectare on average, and only 3,500 to 4,000 for the best plots. Grapes from the various vineyard plots, from the greatest to the least, are fermented separately. Almost from the time the grapes are picked, Mr. Pontallier and his team have some idea of which vats will go into the first wine, Château Margaux, and which into the second wine, Pavillon Rouge. But nothing is certain.

"We taste," he said. "We taste, we taste, we taste."

The new wine would go into 225-gallon barrels in October or early November. When the final selection and blending were done early in 1996, the cut would be severe. As much as 50 percent of each vintage goes into Pavillon Rouge. Under previous owners, only about 15 percent was held out for the second wine.

The 1995 harvest produced 1,600 barrels of wine, which translates to 435,000 bottles, or 40,000 cases. "Because we haven't made our final selection, we don't know how much will go into the first wine, and how much into the second," Mr. Pontallier said, "but it will be around 20,000 cases of each."

In 1977 André Mentzelopoulos, a Paris financier born in Greece, bought Château Margaux from an old Bordeaux wine family, the Ginestets, for $15 million and began a program of renovation still under way. He died in 1980, and ownership passed to his wife and daughter and then to an Agnelli family trust called the Exor Group. Corinne Mentzelopoulos, André's daughter, is on the Exor board and is managing partner of Château Margaux.

The Margaux pickers' party, the *gerbaude,* which celebrates the end of the harvest with food, wine, and song, was held on October 3. "The name predates French," Mr. Pontallier said. "It goes back to a time when people were happy that the harvest, which meant their food—not just wine—was safely stored for the winter months ahead."

OCTOBER, 1995

## SHARING THE RITES OF THE WINE HARVEST

Vineyards and wineries are fascinating all year, but they are at their best during the harvest—those exciting ten days or so in autumn when the grapes are picked and the wine is made. It's called the *vendange* in France, the *vendemmia* in Italy, and in this country, the crush.

There is the earthy smell of the soil and the vines and the newly picked grapes in the crisp autumn air, the sweet-sharp taste of fresh juice, and the undulating rows of vines blazing orange, gold, and red against the distant hills. There is the tension in the vat house as the tumultuous fermentation begins, and when the work is over and the wine is quiet in the barrels, there are the joyous harvest festivals, not much changed from the time when Etruscans sketched them on the walls of Umbrian caves.

The harvest is a celebration of the splendors of nature, one of the simplest and most profound of all human rites. An opportunity to share in it should not be missed.

"I'd love to, but aren't the wineries too busy at that time of year?" Yes and no. A small winery, where the owner is the winemaker, vineyard manager, tractor operator, and cook for the hired hands, may not be the place to go when the grapes are coming in. With larger wineries, it's no problem. Many encourage visitors. They operate just as efficiently with a hundred guests watching as they do when no one is around but the workers.

Visitors who turn up at harvest time are, in fact, the only ones who ever really see the wineries in full operation. A tour guide can explain how a crusher works and what the stemmer does and how pressure builds up in a rotary press, but seeing these exotic machines at work is much more fun. And the only time to do it is when the wine is made, which is just once a year.

In most parts of the world, this means selecting one's time carefully, because most wineries produce but a single wine and it doesn't take all that long. In California a winery may make five or ten wines, with the various kinds of grapes maturing over a period of six weeks or more. Accordingly, a visitor has that much more opportunity to see how a California winery works than one in, say, France's Loire Valley.

Among the larger wineries that can accommodate visitors even during the annual crush are Domaine Chandon, Beaulieu, Franciscan, Robert Mondavi, Christian Brothers, and Sterling in the Napa Valley. Over the mountain in Sonoma, Sebastiani, Buena Vista, Sonoma Vineyards, and Piper Heidsieck are among those worth considering.

In France, harvest time is celebrated according to the style of the wine region: with verve and gusto in earthy Burgundy; with elegance and style in formal Bordeaux. Visitors might find it more difficult to share in the fun in Burgundy simply because most of the wineries are so small. They receive visitors in calmer seasons. During harvest, they are preoccupied with grapes and grape pickers, usually high-spirited university students. The excitement of the Burgundian harvest is infectious, and then there is always the weather, the countryside, and some of the best restaurants in France.

For Burgundians, the highlight of the wine year comes two months after the harvest, at the famous auction at the Hospice de Beaune. Wine buyers from around the world descend on Beaune, the medieval capital of Burgundy, to bid on the wines of the recent harvest. There are parties everywhere, the most famous being those held in the former abbey known as the Clos de Vougeot, where five hundred people sit down to eat dinner, drink great wines, and sing, literally, the praises of Burgundy.

In the Beaujolais region in southern Burgundy, the harvest is not half so exciting as the day, of the release of the Beaujolais nouveau. This is the wine made only six or eight weeks after the harvest and proclaimed in restaurants from Lyons to Los Angeles. There is dancing in the streets when the first nouveau arrives, and this is one wine festival that tourists are particularly welcome to join in. The best way to celebrate the arrival is to stake out a spot at a good bistro in Lyons or Paris and hold your glass out when the first barrel is opened.

Each year, the highlight of the Bordeaux harvest is the Ban de Vendange, a gala party held in the cellars of one of the famous chateaus. Most of the chateaus and many of the big wine companies take tables at the Ban. Many hold their own parties on the days before or after the event itself. An invitation to the Ban is much prized because the chateaus outdo each other in throwing the most talked-about parties and pouring the rarest wines.

The casual visitor may not be able to crack the Ban de Vendange but he or she will find hospitality in many cellars in the region, particularly in St.-Émilion, where the festivities are more casual than in the Médoc. Anyone with specific chateaus on his list should let them know in advance.

Planning is a particularly good idea where wine travel is concerned, if only because winemaking is mostly a small-scale, artisanal undertaking. There are still opportunities to get to know the people who make the product one uses.

SEPTEMBER, 1982

## TIPS ON VISITING VINEYARDS
### IN CALIFORNIA AND FRANCE

Napa-Sonoma harvest time in the California vineyards attracts visitors the way ripe grapes lure starlings. In mid-September, if that's when the grapes are ready, Route 29, the main road through the Napa Valley, is clogged with tourists' cars from early morning to dusk, when the last winery tours are ended. Over in the Sonoma Valley, the crowds are almost as thick. The wise traveler books in advance. The chic little bed-and-breakfast places that have proliferated throughout the valley, such as the Burgundy House and the Bordeaux House, both in Yountville, often are booked months ahead. Even large places such as the lush Silverado Country Club and the simpler Holiday Inn at Napa are tough to get into during the crush. Many visitors simply stay in San Francisco and drive up for the day. The city of Napa and the town of Sonoma are both about an hour's drive north of the Golden Gate Bridge.

To get more information on what's doing in the Burgundy wine country, contact the French National Tourist Office or the Chamber of Commerce in Beaune, France. They can provide information on cellar visits, hotels and restaurants, and, if you are interested, on barge trips and even hot-air balloon trips over the Burgundian vineyards.

For the longest time, the Bordeaux wine people were not interested in tourism. They entertained members of the trade, importers and distributors and even retailers of wine, but not casual visitors. That has changed—not entirely, but enough to make a visit very worthwhile. There are now, for the first time, wine country trips of from one to four days, including weekend jaunts that include stops at chateaus and good restaurants. Again, the French National Tourist Office as well as the Office of Tourism in Bordeaux are good sources for travel information.

## THE RITES OF VINTAGE ASSEMBLY

On a chill, rainy day in early spring of 1983, seven men gathered in the refectory of Château Prieuré-Lichine in the tiny village of Cantenac, twenty miles north of Bordeaux. They had come at the invitation of the owner, Alexis Lichine, to help him make a wine.

Awaiting them on a huge oak table in what used to be the monks' dining room were some two hundred glasses and twenty-two bottles of wine, samples from all the vineyards that are part of Prieuré-Lichine. Their task: to help Mr. Lichine fix the blend that would become his 1982 vintage.

The cycle of the vine—and the wine—begins with the budding in the spring, follows through the flowering of the vines in early summer, and goes on to the harvest and the grape crush in the fall. Each step is important, but none more than the assembly of the wine—the *assemblage*—some six months after the wine is made. It is the day when professional wine men bring all their skills to bear, the day they create a vintage.

The members of the group that morning were Émile Peynaud, director emeritus of the Enological Laboratory at the University of Bordeaux and probably the world's most prominent enologist; Jacques Boissenot, a professional enologist and a specialist in the wines of the Médoc region; Patrick Léon, president of the Enologists of Southwest France and, in his youth, a student and employee of Mr. Lichine; Jean Delmas, the director of Château

Haut-Brion; Hugues Lawton, a Bordeaux wine merchant; Sacha Lichine, a Boston wine importer and Mr. Lichine's son; Edmond Caubraque, the business manager of Prieuré-Lichine; and Mr. Lichine himself.

I had thought that a chateau's vines were all treated the same. I knew there were different kinds of grapes—cabernet sauvignon, cabernet franc, merlot, malbec, petit verdot—but I thought that each property's blend varied little from year to year. And I thought everything was mixed together at harvest time. Well, that's not the way it works.

Prieuré-Lichine was once actually the priory of the Benedictine church in Cantenac. When Alexis Lichine bought it in 1951, it had been a wine property for several hundred years and was known as Château Cantenac-Prieuré. It was a run-down, ragtag, has-been vineyard that had not been properly tended for half a century. This is not uncommon in the Médoc, where owners die and heirs are indifferent. Absentee ownership, inept management, incompetent advice—all contribute to the deterioration and sometimes the destruction of once-proud wine properties. Some Bordeaux vineyards, famous a century ago, exist today in name only, in old books.

Many Bordeaux properties were decimated by legacies. Heirs divided and sold their parcels until some vineyards were crazy quilts of divided ownership. When Mr. Lichine arrived at the Prieuré, he found himself the owner of a jumble of vineyard patches and strips: a few vines here, a few vines there, a small good vineyard here, a big poor one there. He spent the ensuing four decades wheeling, dealing, begging, borrowing, threatening, and cajoling in an unceasing effort to create the kind of vineyard he thought the Prieuré should be.

The key is this: The town of Cantenac is adjacent to the town of Margaux, one of the most famous of all the French wine communes. Cantenac actually shares the right to use the Margaux name, or *appellation,* on its wine. So do several other small villages. As long as Mr. Lichine bought within the confines of the wine area of Margaux, his vines were legitimately part of the vineyard of Prieuré-Lichine.

Sharing the Margaux appellation is an honor. It is also confusing. It's confusing because Margaux is not only an appellation area, it is the name of a town and of a famous chateau within that town, Château Margaux.

Five communes share parts of the Margaux appellation: Margaux itself; Cantenac, where the Prieuré is located; Labarde; Arsac; and Soussans.

Prieuré-Lichine owns pieces of vineyard in each of these communes. It also owns some vines in places outside the appellation area. Wine made from these vines can never be called Château Prieuré-Lichine; its appellation is only Haut-Médoc.

After decades of trading, Mr. Lichine increased the size of the Prieuré vineyard from about twenty-seven acres in 1952 to 143 acres in 1983. In 1982, there were almost 157 acres. He gave up 15 acres in a trade that netted him a small piece of better ground. Better ground is invariably higher ground, with its better drainage. Mr. Lichine was most proud of finally having assembled the land north of the chateau, which he called the knoll of Cantenac. "It's some of the best vineyard land in the Margaux appellation," he exulted. Most of the Bordeaux chateaus engage in this bartering game to improve the quality of their holdings and to consolidate them, to group as many vines as possible around the chateau itself and, failing that, to create large sections of vineyard that are easier to work with mechanical equipment. Almost all new vines planted in the Médoc are laid out with the eventual use of mechanical pickers in mind.

Mr. Lichine's horse-trading netted him pieces of some very famous neighbors. Parts of the Prieuré vineyard once belonged to chateaus Palmer, Ferrière (one of the chateaus that have disappeared), Kirwan, Giscours, d'Issan, and Boyd-Cantenac, all third growths, as well as chateaus Durfort-Vivens and Brane-Cantenac, both second growths.

In the now-famous classification of the Médoc vineyards made in 1855, chateaus were designated first, second, third, fourth, and fifth growths. The Prieuré was made a fourth growth. Mr. Lichine thinks it should be higher now, not only because half of its vineyards were once higher, but also because he has done so much to improve the winemaking. His chances of being upgraded are slim. It took Philippe de Rothschild fifty years to have his Château Mouton elevated from second growth to first.

Once pleasantries had been exchanged that morning, everyone got down to serious tasting. The wines in front of them all had been drawn earlier that morning from the glass-lined concrete vats in which the wine had rested over the winter.

The most important grapes in Bordeaux are cabernet sauvignon, merlot, cabernet franc, petit verdot, and malbec. At Prieuré-Lichine, cabernet predominates, but the blend includes merlot and some petit verdot. When the grapes are crushed in the fall, every effort is made to isolate the wine from the various parts of the vineyard and from the three varieties of grape. To do it perfectly would require more fermenting vats than most medium-sized chateaus can afford, so some merlot is dumped in with some cabernet. And some cabernets are combined.

For example, vat 18 held all petit verdot, but from five separate sections of vineyard: Degas, Coudert, Belloc, Caneron, and an area called 4,000 Lignes. Vat 7 held cabernet and merlot from plots named Blancard, Plateau-Ferrière, and Muraillade. Several vats held only press wine, which is made from the relatively harsh and tannic second pressing of grapes. This wine can be used to increase color or tannin in relatively weak wines.

After about two hours of quiet tasting, sniffing, holding glasses up to the windows, and scribbling notes, the experts tallied up their votes and the best vats were chosen. The 1982 Prieuré-Lichine had been fixed. The 1982 crop was a big one and a very good one. Most of the wine was high quality. Even so, only about two-thirds of the crop would go into the Prieuré blend. There is more to all this than enology. Juggling prices and market conditions in his head, a chateau owner has to decide how he will sell his wine. If he adds too much mediocre wine to the blend, quality suffers. If he's too exclusive and uses only his very best, the wine won't cover its costs.

Thus about a third of the Prieuré's 1982 wine would appear under Mr. Lichine's second label, Château Clairfont. (There actually is a Château Clairfont—the Prieuré manager sometimes lives there—and it is entitled to the Margaux appellation.)

There is also a third label, Campion, which is the name of a portion of the Prieuré vineyards. It is used from time to time for superproduction— that is, wine over the quantity permitted by law in a given year. To keep winemakers from pushing production and turning out large quantities of thin wine—a persistent problem in Burgundy—the quasi-governmental Institut National des Appellations d'Origine Controllée decides each year just how much wine can be produced per acre of vines. Anything over that must be sold as inexpensive local wine.

In 1982 the I.N.A.O. granted a one-time-only increase, enabling Mr. Lichine to produce almost 90,000 gallons of appellation wine. His normal production was about 55,000 gallons.

The *assemblage* of the 1982 was a success. It must have been: At the lunch afterward, Mr. Lichine served a good Burgundy—heresy in staid Bordeaux.

JUNE, 1983

## WHERE VINES STAIR-STEP
## UP THE HILLS

Over the years, I'd tasted and enjoyed a few bottles from a California winery called Renaissance. I'd even noticed the odd appellation: North Yuba County. But I never bothered to look it up.

Then I read James Halliday's comments on Renaissance in his excellent *Wine Atlas of California* (Viking, 1993). "If there is a more remarkable vineyard in California," he wrote, "I did not see it. Those who have visited the Douro in Portugal or gazed upon the hill of Hermitage in the Rhône Valley will understand the impact Renaissance has on the first-time visitor."

He continues: "How did those Portuguese construct those endless terraces, so far above the River Douro, so far from summer water and so far from civilization as we know it today, let alone 300 years ago? The same sense of the improbable, accentuated by the grandeur of the sweeping vistas, confronts one at Renaissance."

Mr. Halliday explained that the vineyard and the winery are owned by a group—some people would say a cult—called the Fellowship of Friends, founded in Berkeley, California, in the early 1960s, and he concluded, a bit breathlessly, with these two sentences:

"Renaissance Winery is open to visits by appointment only. I can only suggest you move heaven and earth to make an appointment, for you will see both when you arrive."

So, intrigued by Mr. Halliday's enthusiasm and armed with a map faxed to me by Joseph Granados, the Renaissance sales and marketing vice president, I set out from Sacramento on the eighty-five-mile drive to Oregon House, a village in the remote foothills of the Sierra Nevada, home to the winery and vineyard as well as the Fellowship's headquarters.

Mr. Halliday had not exaggerated. The Fellowship of Friends owns 1,400 acres of what is mostly wilderness, but 365 of those acres have been converted into a spectacular vineyard. At an elevation of 2,300 feet, it is probably the largest mountain vineyard in North America.

More than anything else, it resembles an immense bowl formed by the surrounding hills. The vines are planted on terraces—almost a hundred miles of them—lining the inner walls of the bowl. So steep are some of the slopes that almost all planting, pruning, and harvesting must be done by hand. Grant Ramey, the vineyard manager, said that temperatures can vary ten degrees between the top and bottom of the more precipitous inclines.

In most large vineyards, grapes are planted in blocks of several acres, delineated by soil types, drainage, and exposure, among other things. At Renaissance, when the winemaker, Gideon Beinstock, makes his blends, he talks of wines from slopes. And the slopes are numbered, from 1 to 27.

The Fellowship of Friends calls the entire property Apollo. They bought it in 1971, mostly because Yuba is among California's poorest counties and the land was cheap. But also because their leader, a former schoolteacher named Robert Earl Burton, had said that only Apollo would survive the nuclear holocaust he has predicted for 2006.

Beginning in the early 1970s, members of the Fellowship of Friends, who must devote at least a month a year to the group's activities, assembled from all over the world and carved the vineyard out of the rocky soil, literally by hand. They pulled out pine, cedar, scrub oak, and manzanita. They dynamited and hauled away hundreds of tons of granite boulders, then spent four years building the terraces.

The first vines were planted in 1975, but not until 175,000 holes had been drilled into the granite base and filled with compost. Planting continued until 1983, and additional vineyard sites have been selected for future development.

About 50 percent of the vines are cabernet sauvignon. Much of the rest of the vineyard is planted with riesling and sauvignon blanc, but there are substantial blocks of chardonnay and merlot, and smaller quantities of syrah, sangiovese, mourvèdre, grenache, pinot noir, petit verdot, and viognier. Mr. Beinstock is also experimenting with Portuguese varieties used in making Port.

Renaissance's first winemaker was Karl Werner, who had once made wine at Schloss Vollrads in Germany and later at Callaway Vineyard and Winery at Temecula, south of Los Angeles. The importance of riesling at Renaissance is traceable to Mr. Werner's influence. For a time after his death in 1988, his wife, Diana, served as winemaker.

Mr. Beinstock, who describes himself as French-Israeli, took over in

1993. He had arrived at Renaissance in the 1970s and helped clear the land. After spending most of the 1980s in France, he returned to Renaissance as an assistant winemaker. He took over in time to make the 1993 vintage and was given the title of winemaker in January 1994. He soon shifted the winery's focus from German-style white wines to Bordeaux-style reds and quickly won critical acclaim for both the 1993 and 1994 cabernet sauvignons.

The first commercial vintage at Renaissance was in 1988. In a departure from normal marketing strategy, the winery worked to develop an overseas market first. "The American market was weak in the late 1980s, and we weren't known," said Mr. Granados, the vice president. "We decided to go where we would be on a more even playing field." Today Renaissance wines can be found in Europe and the Far East. "There is some in Israel and even a bit in Russia," Mr. Granados said.

Perched strikingly on a hill amid the vineyards is the Renaissance winery, which, as Mr. Halliday said, is "functionally complete." Inside it's a spotless state-of-the-art winery; outside, it's a raw concrete structure, its only decoration the rusted steel reinforcement rods jutting from the walls. One day it will be finished, Renaissance people said, but since there is already a reception center and tasting room, an attractive small restaurant, and a gift shop on the grounds, there seems to be no sense of urgency about making the winery more attractive.

Besides, the money is needed elsewhere. Renaissance produces about 35,000 cases of wine annually and is only beginning to show a profit, Mr. Granados said.

Although Renaissance is a Fellowship of Friends project, it is a separate for-profit entity with its own board. The Fellowship was viewed with some concern locally when its members first arrived in Oregon House, but they have come to be regarded as good neighbors. Although the Fellowship itself is tax-exempt, the Renaissance winery is Yuba County's third-largest taxpayer, according to the county assessor's office.

The Fellowship claims about 1,900 members worldwide, some 600 of whom live at or near the Apollo property. Others live and work at some sixty-five teaching centers around the world. They believe that intense self-awareness, a positive outlook, and a commitment to art and high culture lead the way to a higher consciousness. And they see wine as part of that commitment.

"Wine symbolizes our art of living," a Fellowship brochure says, "and is our labor of love."

FEBRUARY, 1997

## AT RIDGE VINEYARDS

In my twenty-five years at Ridge Vineyards," Paul Draper said, "I've made thirty different single-vineyard zinfandels."

To a dedicated zinfandel fan or another winemaker, that's a remarkable statement; anyone else will require a little explanation.

In France, particularly in Bordeaux, there is usually one name and one wine. Château Latour, for example, makes a famous red wine from its own vineyard adjoining the chateau. And that's all it makes. It may differ from vintage to vintage, but it is always Château Latour.

Ridge Vineyards is situated high in the Santa Cruz Mountains south of San Francisco. There is no chateau, of course—just some ramshackle sheds and buildings that convey a deceptive air of casual rusticity. Like Latour, Ridge has a vineyard, and like Latour it produces fine red wine.

There the resemblance between Ridge and a traditional winemaking property, in this country or France, ends. Ridge has never confined itself to home. Almost since its first vintage in 1962, Ridge's winemakers—there have been only two—have scoured the state of California, seeking out the best grapes and turning them into wine.

Ridge is not alone in ranging far afield. California vintners are a restless lot, convinced that the perfect vines are just over the next hill. Some buy grapes to enhance wines made from their own vineyards. Others will make and sell three or four versions of the same wine, using grapes from a different vineyard for each.

But none has taken the purchased grape concept as far as Ridge. Over the years, Ridge cabernet sauvignons have been made from the home vineyard, Monte Bello; from the Jimsomare Ranch on the same mountain; from a blend of the two, labeled Santa Cruz; from the York Creek Vineyard on Spring Mountain in Napa; from the Tepusquet Vineyard in the Santa Maria Valley; and from the Beatty Ranch, on Howell Mountain in Napa, among others.

There is a Ridge cabernet style: voluptuous and immensely concentrated with complex layers of flavors including black currant and cherry, a dark garnet color, and a rich, lingering bouquet reminiscent of cedar and smoke. Each Ridge cabernet expresses the Ridge style in its own way, although few have ever reached the heights of the best Monte Bellos. No one knows how long these wines will last. The very first vintage of Monte Bello, 1962, is still a superb wine, according to people who have recently tasted it. Because they are cheaper, Ridge zinfandels are better known than the cabernets. They start at around $20; Monte Bello is in the $100 range. There are also many more zinfandels, as Mr. Draper, Ridge's winemaker and chief executive, indicated.

The first Ridge zinfandel was made in 1964 from a vineyard, the Picchetti Ranch, near the winery. In 1966 the first Geyserville zinfandel was made. It came from grapes from the Trentadue Winery in Geyserville, in northern Sonoma. The Lytton Springs Vineyard in the Dry Creek area of Sonoma became a regular source. Ridge now owns the vineyard. Other zinfandels came from a variety of sources in Amador County, including vineyards in Fiddletown, the Shenandoah Valley, and the Amador Foothills.

The York Creek Vineyard has been an important source; other zinfandels have come from Howell Mountain, the Alexander Valley, and Paso Robles in San Luis Obispo County.

Currently, Ridge Vineyards produces three cabernet sauvignons, Monte Bello, York Creek, and Santa Cruz Mountains; one chardonnay, from the Santa Cruz Mountains; and a petite sirah and a merlot, both from York Creek.

There are six zinfandels, four from Sonoma County. These include one from the Pagani Vineyard in the Sonoma Valley, one from Lytton Springs in Dry Creek, one from Geyserville, and one simply designated Sonoma that is a blend of grapes from the first three that don't meet Mr. Draper's standards for single-vineyard wines. No Sonoma Zinfandel was made in 1992 because grape quality was so high. The fifth zinfandel comes from Paso Robles in San Luis Obispo, and the sixth from York Creek in Napa.

Mourvèdre, a grape usually associated with the Rhône Valley in France, has recently become popular in California, where a few vines of it have long been cultivated under its Spanish name, *mataró*. Ridge now makes a mourvèdre but calls it "mataró" because, Mr. Draper said, "an American winery should stick to traditional American names." The grapes for this wine come from a vineyard in Oakley, in Contra Costa County.

In recent years, Ridge has begun to return to its roots, at least for some wines. Chardonnay was made from Howell Mountain and Santa Cruz grapes in 1991. In 1992 it was all Santa Cruz. Merlot, originally made from Bradford Mountain (Napa) grapes, is now being made entirely from Santa Cruz grapes. And, Mr. Draper says, Santa Cruz cabernet is slowly getting better than the much-praised York Creek.

York Creek cabernet was declassified in 1988 and 1989 because Mr. Draper didn't think it was up to his standards. It was sold as Napa Valley cabernet. The 1991 was the last vintage under the York Creek name. "We have to work extremely hard to make some of our wines," Mr. Draper said, "and York Creek cabernet was one of them. More and more we're finding that wines from the home vineyards seem to make themselves."

Ridge quality rarely goes unnoticed. Many Ridge wines are sold out soon after release. Quantities are small. Only 3,500 cases were made of the Monte Bello currently on sale, 1989. Production of York Creek 1991 was 1,374 cases; of Santa Cruz Mountains, 7,400 cases.

Current zinfandel production is Geyserville, 8,000 cases; Lytton Springs, about 12,000; Pagani, 4,000; Paso Robles, about 5,000; and York Creek, about 3,500. Merlot, chardonnay, and mataro bring the total to about 40,000 cases a year.

JUNE, 1994

## TWO AND A HALF DECADES LATER, A VERY GOOD YEAR

They celebrated a milestone at Opus One in August 1994. Opus One is the name of a winery, an architectural tour de force in the Napa Valley in California; it's also the name of its wine.

The celebration was for the 1991 vintage, the first to be made at the winery. But the party was almost twenty-five years in the making, having survived big egos, a big price on the wine, a big construction problem, and a serious disease in the vines.

The idea for Opus One was born at a wine and spirits industry convention in Hawaii in 1970 when Baron Philippe de Rothschild casually proposed it to Robert Mondavi, the California winemaker. Baron Philippe was the owner of Château Mouton-Rothschild at Pauillac, near Bordeaux, one of the most famous wine properties in France. Mr. Mondavi's winery in the Napa Valley had opened to almost instant acclaim only four years earlier.

The two men met again in 1978, at Château Mouton-Rothschild. After twenty-five minutes of discussion, they agreed on the outlines of a joint winemaking venture. They would create a California wine combining the best of two worlds, or more precisely, of two wine regions, Bordeaux and the Napa Valley.

The two men had much in common, even though one was the son of poor Italian immigrants and the other a scion of one of the wealthiest families in Europe. Both had been athletes in their youth. Both were restless innovators. Both were clever showmen possessed of outsize egos.

Baron Philippe loved California. Briefly during the 1930s, he had dabbled in Hollywood as a producer. After World War II, he and his wife, Pauline Fairfax Potter, an American, vacationed regularly in Santa Barbara. In the early 1970s, the French economy was more unstable than usual, and the booming California wine business seemed like a most prudent investment.

For the visionary Mr. Mondavi, the French connection offered a partner with deep pockets and a measure of prestige that was undreamed of in the days when, as a boy, he had nailed together crates for shipping cheap grapes to the East Coast.

The project moved along rapidly. Even the unusual name of the red wine was chosen quickly.

According to one version of the event, Baron Philippe called from Paris and suggested Opus. Mr. Mondavi demurred; it was too musical. Two days later the Baron called back. "How about Opus One?" he said. And that was it.

Until a separate winery was built, Opus One would be made from Mondavi grapes at the Mondavi winery in Oakville. Mouton-Rothschild would supply its winemakers and oak barrels for aging and would underwrite other costs. The first vintage was in 1979. The Opus One telephone number commemorates that year: 963-1979.

Thanks to a carefully orchestrated promotional campaign, the 1979 vintage was in demand almost from the time the grapes were picked. One case went for $24,000 at a charity auction three years before the wine was sold commercially.

When the 1979 vintage was released in 1984—simultaneously with the

1980 vintage—it was obvious that a star had been born, even though it cost $50 a bottle, which was then a startling price. Opus One was truly a California wine with a French pedigree.

That same year, 1984, the Mondavis and the Rothschilds chose Johnson, Fain & Pereira Associates, architects in Los Angeles, to design their winery. By then, Baron Philippe's daughter, Philippine, was leading the French half of the Opus One team on behalf of her aging father, then eighty-two. (The Baron died in 1988.)

The architects drew up plans for a building, across the highway from the Robert Mondavi Winery, that would have eighty thousand square feet of space, of which fifteen thousand square feet would be reserved for a huge curved barrel room. Throughout the winemaking process, an ingenious gravity-flow system would eliminate the need to pump juice or wine from one level to another. The winery was budgeted at $10 million. Construction started in July 1989 and was finished in December 1991.

Then misfortune struck. Twice.

The plan was to bury much of the winery below ground, for natural cooling. The engineers discovered, too late, that the building was sitting on an exceptionally high water table. What's more, it was found to be close to geothermal springs that raised the ground temperature to about 72 degrees. An expensive cooling system had to be devised and installed.

Meanwhile, a new phylloxera epidemic had begun to destroy California's vineyards. Phylloxera is an aphid that feeds on grapevine roots. In the late 19th century, the bug, which originated in this country, wiped out most of the world's vineyards. Only the discovery of phylloxera-resistant rootstocks enabled the wine business to survive. Since the bug works only underground, the vulnerable vines can be grafted, above ground, to the resistant roots.

A new strain of the bug, which attacks rootstocks once considered impervious to it, has forced extensive replantings in California. In 1990 forty of Opus One's seventy-five acres of vines were found to be infected and had to be replaced. Thus, fifteen years after its first vintage, the wine was still being made with grapes from vineyards owned or leased by the Mondavi winery. Today Opus One is made entirely from its own grapes.

New construction and replanting increased the cost of the Opus One winery. The winery's management said the final cost was $15 million, although industry estimates have put it as high as $25 million.

The winery, which opened in 1991, has been variously described as a space station, a Mayan temple, and a futuristic mausoleum. In fact, it's a strikingly handsome structure built from white Texas limestone and untreated redwood and filled with imported antiques. Its elegance, inside and out, reflects the elegance of its wine and to some extent perhaps the almost casual excess of the mid-1980s.

The wine, a blend predominantly of cabernet sauvignon, with cabernet franc and merlot, has maintained its quality and popularity through the vintages released to date. It is one of the most expensive wines produced in this country: its recent vintages sell for about $65 a bottle, and older ones, when they turn up, can bring $200 or more.

Production began at about 12,000 cases in 1979. Some 27,000 cases of the 1991 vintage were bottled last year, according to winery figures. Production should eventually level off at 30,000 cases, Mr. Mondavi said.

Opus One sells around the world—except, it seems, in France.

"The French are amused by Opus One," Baroness Philippine said. "They're interested, but they're not for the moment prepared to buy it."

AUGUST, 1994

# HOPPING THE GLOBE TO TEND THE GRAPES

There are lots of flying winemakers, but only one is supersonic. His name is Michel Rolland, and he lives here in Pomerol, one of the more famous wine communes in the Bordeaux region. That is, he lives here now and then.

A flying winemaker is an enologist who, for a fee, will go anywhere in the world to advise another winemaker. It might be a short "fix this, fix that, give me a call if it doesn't work" relationship, or it can be a long-term arrangement, with the consultant taking charge of winemaking for several years.

Mr. Rolland, a compact forty-seven-year-old with dark hair, an infectious grin, and a penchant for well-tailored suits and natty ties, has both kinds of relationships in every part of the world where wine grapes will grow. "I have about, oh, a hundred clients," he said. "Some come, some go, the list changes all the time." Mostly it grows longer.

Many winemakers are enologists or, at least, have studied enology in school. And many enologists make wine, including Mr. Rolland, who owns four chateaus here and in neighboring St.-Émilion.

But in recent years the term *enologist* has come to refer to men and women who confine their activities to advising others. The first modern enologist was Louis Pasteur. He spent only a few years of his life on wine, but he discovered how fermentation comes about and what role yeasts play in changing grape juice into wine. The most famous contemporary enologist is Émile Peynaud, now in retirement, who taught at the University of Bordeaux and consulted in France, Spain, and occasionally in the United States.

Pasteur rarely left his laboratory; Peynaud rarely left Bordeaux. Michel returns to Bordeaux only to see local clients between trips. "Last year I made one hundred sixty-seven flights," he said. He has clients in France, many of them only minutes away. Others are in Italy, Spain, Germany, Morocco, Greece, South Africa, Chile, Argentina (where he owns another vineyard), the United States, and even India, where a wine industry is just getting started. Among his newest winery clients in America, all in California, are Araujo, Staglin, Franciscan, and St Supery, all in the Napa Valley, and Simi, in Northern Sonoma.

Historically, winemaking encompasses two disciplines, enology and viticulture, which confines itself to the vineyard and grape growing. But more and more, as winemakers come to understand how important the vineyard is, enologists like Mr. Rolland involve themselves in grape growing. "The most important factor in the making of great wine is ripe fruit," he said, adding, "In areas like Bordeaux and Burgundy, ripe fruit is not always easy to obtain. A succession of mediocre vintages in Bordeaux in the 1960s made it clear, even if it was not too obvious at the time, that it's impossible to make good wine from immature grapes." Is there a Rolland method? "My first recipe is that I don't have a recipe," he said. "Conditions are different and problems are different from one vineyard to another, from one part of a vineyard to another."

Even so, there are Rolland trademarks. He is a strong proponent of two techniques to help grapes reach full maturity: *effeuillage,* which means thinning out the leaves to expose the grapes to sun and air, and *éclairage,* which means reducing the number of grape bunches per vine to lessen the strain on the roots. Also, he advocates using new oak barrels. "Sulphuric acid develops in older barrels," he said. "They say that I am a wood lover, but I use wood to enhance the quality of the wine, not to give it a wood taste. Always, I look for equilibrium in a wine."

Mr. Rolland argues for harvesting as late as possible and for long macerations to enhance color, intensity, and aging ability. Not everyone agrees. Christian Moueix, whose family owns several chateaus in the same region, including the nonpareil Petrus, opts for early picking and shorter maceration for the area's dominant merlot grape, to give the wine more acidity and elegance.

Some winemakers compare Rolland's style with the up-front, full-bodied wines of California. It has even been said around Bordeaux that hiring Rolland is a good way to get a good rating from Robert M. Parker Jr., the wine critic, who has always preferred the richer, softer style.

"There are no secrets to what I do," Mr. Rolland said. "I make wine the way I made it twenty years ago, but always with small changes. Yes, there are still winemakers who say, 'I make wine like my father and his father before him,' but you can be sure of it, those are the vineyards that are going downhill. The best way to negate quality is to ignore the technological developments and research findings of our time."

He practices what he preaches at his own properties, the chateaus Le Bon Pasteur in Pomerol, Bertineau-St. Vincent in Lalande de Pomerol, Fontenil in Canon-Fronsac and Rolland-Maillet in St.-Émilion. He may have no secrets but he does have a secret weapon: his wife, Dany, who is also an enologist. When Michel is on the road or in the air, she looks after the dozens of local clients, including the many small growers who patronize the Rollands' busy laboratory in Pomerol.

His workload notwithstanding, Mr. Rolland shows no signs of fatigue. "I like traveling," he said. "I like seeing new places and hearing new ideas and learning new things." And he is modest about his reputation. "There are no great enologists," he said. "Only good grapes."

OCTOBER, 2000

## THE DAY CALIFORNIA SHOOK
## THE WORLD

On May 26, 1976, six weeks before America's bicentennial celebration, there occurred in Paris a tasting that American winemakers and cultural historians have come to characterize as a defining moment in the evolution of fine wine in this country.

The effects on the California wine industry, as well as on its distributors and their customers, were profound. The winemakers had known they could make great wine; wineries like Beaulieu and Inglenook had been doing it for many years. What the Paris tasting did was bolster American self-esteem. It encouraged many who had been content making mediocre wine to go for it, to make the very best. More important, it showed American consumers that they no longer had to look abroad for fine wine. Little more than a decade earlier, a general American embarrassment had greeted President Lyndon B. Johnson's decree that American embassies serve only American wines.

At that now-famous tasting twenty-five years ago, nine judges, all prominent French food and wine specialists, gathered in the enclosed courtyard of the Paris Intercontinental Hotel and tasted twenty wines, red and white, French and American. The tasting was blind; all the bottles were covered; no one was told which were French, which were American.

When the bottles were unwrapped, the tasters were astonished to discover that the highest scorers were American: a 1973 cabernet sauvignon from Stag's Leap Wine Cellars and a 1973 chardonnay from Chateau Montelena, both from the Napa Valley. Under the headline "Judgement of Paris," *Time* magazine said the "unthinkable" had happened: "California defeated all Gaul."

The tasting had been arranged by Steven Spurrier, a thirty-four-year-old Englishman who ran a wineshop and wine school just off the Place de la Madeleine. The idea sprang from his association with his young American customers, many of whom worked at I.B.M.'s French headquarters just across the street.

In 1975, intrigued by his clients' talk of American wines, Mr. Spurrier sent an assistant, Patricia Gallagher, to California to investigate. She was so impressed that Mr. Spurrier soon made his own pilgrimage westward. He returned to France excited by what he had found and determined to arrange a tasting event. He hoped to twit the xenophobic French, publicize the fine California wines he had discovered, and, understandably, win some publicity for himself.

He succeeded on all three counts. The French tasters, who included the owners of two famous Paris restaurants, the sommelier of a third, and the editor of France's best-known wine magazine, were true to form. They lavished praise on the wines they thought were French—and derided those they assumed to be American.

The California reds were all cabernet sauvignon; the French reds all cabernet-based Bordeaux. The California whites were chardonnay; the French whites all Burgundies, all chardonnays as well. The reds, in order of finish, were: Stag's Leap Wine Cellars, 1973; Mouton-Rothschild, 1970; Haut-Brion, 1970; Montrose, 1970; Ridge Vineyards Monte Bello, 1971; Léoville–Las Cases, 1971; Mayacamas Vineyards, 1971; Clos de Val, 1972; Heitz Cellars Martha's Vineyard, 1970; Freemark Abbey, 1969. The whites, in order of finish, were: Chateau Montelena, 1973; Domaine Roulot, Meursault-Charmes, 1973; Chalone Vineyard, 1974; Spring Mountain-Vineyards, 1973; Joseph Drouhin, Beaune Clos des Mouches, 1973; Freemark Abbey, 1972; Ramonet-Prudhon Bâtard-Montrachet, 1973; Domaine Leflaive, Puligny-Montrachet, Les Pucelles, 1972; Veedercrest Vineyards, 1972, and David Bruce, 1973.

"The egg on the judges' collective faces," wrote Paul Lukacs in *American Vintage,* "came from their inability to discern what until then everyone assumed was obvious—namely, that great French wines tasted better than other wines because they tasted, well, French."

Still, there is one notion that the tasting should have dispelled, but didn't: that California wines do not last. I don't know why the doubt persisted, but, perhaps because it did, Mr. Spurrier decided to recreate the red-wine tasting in 1986. By then the wines were thirteen to sixteen years old. The French wines were in perfect condition, but so were the Americans. And again a California wine placed first—this time Clos de Val—and the Ridge Monte Bello was second. The Stag's Leap 1973 was sixth, after three French wines, but hardly because of age. Warren Winiarski, who owns Stag's Leap Wine Cellars, served it at a dinner I attended at Stag's Leap recently, and at twenty-eight years of age it was in beautiful condition.

For the French, the impact of the 1976 tasting and its 1986 re-creation has been subdued. One early result was that it became easier for French and American winemakers to overcome their mutual suspicion and begin to exchange visits and information. It's a rare California winery these days that doesn't employ a French apprentice or two, while young Americans can be found pulling hoses and scrubbing tanks in wineries all over France. One unexpected effect: American wines may still be rare in the French market, but the French wines that come here taste more and more like California wines. Once Bordeaux winemakers strived for restraint, elegance, and subtlety in their wines. They were proud to say their wines took thirty years to mature. Many still are, but many others in recent years have begun to make California-style full-bodied, high-alcohol, extra-ripe wines that they say are ready to drink in three our four years. Could it be a case of "if you can't beat 'em, join 'em"?

One thing that has changed for both countries over the last quarter-century is prices. In its June 7, 1976, article on the Paris tasting, *Time* magazine noted: "The U.S. winners are little known to wine lovers since they are in short supply even in California, and are rather expensive."

*Time's* definition of "rather expensive"? "$6 plus."

MAY, 2001

# CHAPTER 3

# ALL IN
# THE FAMILY

There are some notable wine families in America, the Gallos, the Mondavis, the Benzigers, whom you will meet in these pages, but their histories pale beside those of the great wine families of Europe. At the house of Louis Latour in Burgundy, for example, there really is a Louis Latour. In fact, there have been Louis Latours, seven generations of them in an unbroken line, since the house was founded more than two hundred years ago.

Even two hundred years is a relative brief time. The Trimbachs and the Hugels, in Alsace, have been making wine for more than three hundred years. The Fallers, another Alsatian wine family, have been at it only a hundred years or so, but the house is presently run by three women, who are also profiled here. Then, of course, there are the Antinoris in Tuscany. They have been plying their trade for an astonishing six hundred years and are still among the top innovators in the Italian wine trade.

In Bordeaux, however, many of the old wine families have disappeared. Under French law property must be divided among all their heirs. As a result, many of the great wine properties were owned by dozens of bickering relatives and were easy prey for insurance companies and other corporate investors in the late 1980s and early 1990s. With the notable exception of men like Anthony Barton, few descendants of the Irish who came to Bordeaux in the eighteenth century to make their mark in the wine trade are left, but their legend lives on in the chateaux they built: Lynch-Bages, Phélan Ségur, and, yes, MacCarthy. The Rothschilds have remained, are still powerful, and have even expanded. Baroness Nadine de Rothschild is the newest of the family's winemakers.

The Benzigers are a classic wine family, even if they did all come from New York City just a couple of decades ago. It may be too early to tell, but it looks like they are going to keep wine in the family—and the family in wine—for many years to come.

## MEANWHILE, BACK IN ALSACE

It had been raining for days. The low gray clouds scudded in over the Vosges Mountains to the west and dumped their sodden contents on the vineyards, then pushed eastward over the Rhine and the Black Forest. It was a typical winter day in Alsace.

But at Domaine Weinbach, a handsome but understated chateau—more a lived-in country house, actually—just outside the picture-book medieval town of Kaysersberg, all was warm and inviting. Here in the paneled front parlor, with its heavy, dark furniture and fading family photographs, Laurence Faller was pouring wine. For the last eight years, she has been the winemaker at the domaine, one of the best-known wine estates in Alsace.

Disappearing regularly into the kitchen, she would reappear with yet more bottles of her latest handiwork, the 1998 vintage. For comparison purposes, there were a few 1997s, too, along with a couple of 1999s, which were still mostly fermenting juice.

Domaine Weinbach is famous not only for its wines, but also for the fact that it is run entirely by women. Ms. Faller, thirty-three, is studying enology at Beaune, in Burgundy, and has an M.B.A. and a degree in chemical engineering. Her sister, Catherine, who popped in and out during our two-hour tasting, is forty-four and handles sales and marketing. Their mother, Colette, nominally the president and chief executive, is better described as the dowager queen. Her age? "Do you have to publish that?" Well, no, probably not.

In the best tradition of French wine widows (Clicquot, Pommery, Bollinger), Colette Faller took charge on the death of her husband, Théo, in 1979 and transformed Domaine Weinbach into one of northeastern France's most prominent wine names.

Théo Faller was the winemaker; when he died, Colette took over with the aid of a family friend, Jean Mercky. Photos show her in boots and old sweaters, hard at work in the damp cellars. Laurence joined the business in 1993; Catherine has always been in the business.

Fallers have owned the property here since 1898, but in Alsace they are relative newcomers. Families like the Trimbachs and Hugels had been making wine in the region for three hundred years when the Fallers, tanners in Kaysersberg, got into the business.

Domaine Weinbach was once part of church-owned vineyards. Augustinian monks acquired the land in the 9th century. In 1612, the Augustinians donated a part of their estates to Capuchin Franciscan friars, who built a convent and church on the land and continued to develop its vineyards and farms. The home vineyard, surrounding the house and winery, is known as the Clos des Capucins. *Weinbach,* in the Alsatian dialect, means "wine stream" and is in fact the name of the small stream that flows past the domaine on its way to the Rhine.

An inventory made by the Capuchin friars in the 17th century listed fourteen thousand to fifteen thousand vines, including muscat, "riesselin" (riesling), and "chaselin" (chasselas). The friars also raised snails, and once a year they gave a banquet for the local notables, at which one of the important dishes was *escargots au riesling.* The friars were ousted at the time of the French Revolution, and the property passed through several hands before the Fallers acquired it.

Originally the estate consisted of the chateau and the twelve-acre vineyard within the estate walls, or *clos.* In the years since World War II, the property has been expanded to about 150 acres, some of it leased vineyards. As an owner-harvester, under French wine regulations, the Domaine Weinbach is not allowed to buy grapes from other growers. But the Fallers may rent vineyard property, as long as they grow and harvest the grapes themselves.

Like other French vineyard regions, the best Alsatian vineyards are classified as *grand cru,* which translates—badly—as "great growth," and as *premier cru,* or "first growth." Domaine Weinbach owns two *grand cru* parcels; about forty acres on the Schlossberg, a hilly slope just north of the domaine, and a piece of an adjoining vineyard called the Furstentum. The Schlossberg vines produce the domaine's best riesling; the Furstentum is the source of its best gewürztraminer. Other vineyards, including the Altenbourg, which is unclassified, turn out pinot blanc, pinot gris, and muscat.

The best grape in Alsace is the riesling; the best-known grape is the gewürztraminer. Gewürztraminer produces a pungent, flowery wine with a taste that some experts compare to litchi nuts. Even when fermented completely dry, it can seem to be sweet because of the flowery bouquet. Riesling produces wines that, at their best, display layers of complexity and nuance. A good Alsatian riesling can last fifteen years, and late-harvest versions can live in superb condition for thirty to fifty years.

Alsace produces a lot of wine, most of it indifferent stuff that, fortunately, rarely makes it across the Atlantic. Michel Bettane, the French critic who tasted with me at Domaine Weinbach, contends that in Paris, even some of the *grand cru* wines are watery and sugary. "Less than 10 percent of Alsatian wines are excellent," he said. "The other 90 percent are sugar water."

"Michel," Laurence Faller said, "tends to get carried away."

At the other end of the spectrum, too many of the best Alsatian wines are being made in the sweet style. Some of the finest producers are turning out wines that make fine aperitifs but very poor companions for most meals. Even Ms. Faller acknowledged that all of her 1998 gewürztraminer could have qualified for the *vendange tardive,* or late-harvest, designation. Late-harvest wines are invariably much higher in sugar content than dry table wines.

The total Domaine Weinbach production is about fifteen thousand cases, about half of them exported to twenty-three countries. The United States is the biggest customer. The domaine sells about a quarter of what it makes to tourists, mostly German.

My favorite Faller wines include the Riesling Schlossberg and the Riesling Schlossberg Cuvée Ste. Catherine, Cuvée du Centenaire, both from 1997 and both in the $45 to $65 price range. Perhaps the best introduction to the Weinbach rieslings is the less subtle Réserve Personnelle, which sells for around $24. The 1998 Riesling Cuvée Ste. Catherine is a big, powerful wine that will last and last. It hasn't been shipped yet, and it will be expensive.

The real gems in the Faller collection are the late-harvest wines. The 1995 Pinot Gris Quintessence des Grains Nobles, Cuvée du Centenaire, for example, is made from individually picked late-harvest grapes and sells for around $300 a half bottle.

One of the best ways to enjoy good Alsatian wine is with the local cheese, Muenster. After our tasting, Colette Faller, elegant in jewels and eyeliner, swept in to announce lunch. It was simple fare, chicken and noodles and Muenster, accompanied by some of the best Faller wines. Anyone who says no meal is complete without red wine has never sat down to lunch in the big, warm kitchen at Domaine Weinbach.

JANUARY, 2000

## THEY'RE WEARING THE GREEN AMID THE GREAT REDS

It is often said that Ireland's most important export has been its youth. In the nineteenth century, famine and despair drove well over a million of the Irish to the United States alone.

The enormity of that sad exodus, still vivid to many, tends to blur the memory of earlier waves of Irish emigrants, including the famous Wild Geese, who came to France in the late seventeenth and early eighteenth centuries.

Many of the Irish immigrant families went into the wine business, where their descendants are often still to be found. In the eighteenth and early nineteenth centuries, Ireland imported more fine Bordeaux wine than did England.

"Someone came up with the name 'Wine Geese,'" said Anthony Barton, the urbane doyen of Château Langoa-Barton and Léoville-Barton in St.-Julien. "Of course, it's just a pun. There's no society or club or any such thing," he added, plainly appalled at the idea.

Every year French descendants of those immigrants, usually joined by Ireland's ambassador to France, pay homage to their forebears at a round of St. Patrick's Day celebrations in Bordeaux and in the nearby wine country.

According to legend, the first Irish immigrants to France were soldiers defeated in battle, unbowed Catholic warriors fleeing English tyranny at home to follow the Stuarts into exile and carry on the fight with their old foe. And some were just that. But not all left to join the French army's independent Irish Brigades. Many were merchants, some even Protestants, who came to prosperous France to seek their fortune.

Mr. Barton is a descendant of one of those men, Thomas (French Tom) Barton, who left Fermanagh in 1722. First at Marseilles and then at Montpellier, he established himself as a trader. He arrived in Bordeaux in 1725 and amassed a fortune by trading French brandy and Cognac for Irish wool. Although the Bartons always kept close ties with Ireland—Anthony Barton was born there—since French Tom's arrival, there have been nine generations of Bartons here, an unbroken line.

The Bartons were one of the first of the prominent Irish merchant clans of Bordeaux, but by no means the only one. Indeed, the Irish presence here in the 18th century was formidable. Renagh Holohan, in her book *The Irish Chateaux* (Lilliput Press, Dublin, 1989), says there were some sixty Irish names among the city's merchants in the mid-1800s.

Most of the old Irish families have disappeared. Some moved away; some simply died out. Others had daughters who married into the French aristocracy. Even so, many of their names live on in the great wine chateaus they once owned: Langoa-Barton, Léoville-Barton, Lynch-Bages, Lynch-Moussas, Kirwan, Phélan Ségur, Dillon, Clarke, and MacCarthy among them.

Some Irish families bought chateaus and kept the ancient names. The Johnstons, for example, owned the chateaus Dauzac and Lascombes in Margaux, Ducru-Beaucaillou in St.-Julien, and had shares, with the Bartons, in Latour in Pauillac. Château La Houringue, whose vineyards are part of Château Giscours's vineyard now, was owned by John O'Byrne, a merchant and shipper who was known as the Chevalier O'Byrne.

Hugh Barton, who headed the Barton family business at the time of the Terror, in 1793, was imprisoned in a Carmelite convent, along with clerical students from Bordeaux's Irish College. He was released and fled to Ireland, leaving his business in the hands of his partner, Daniel Guestier. Barton & Guestier, their firm, survived independently until 1956, when it was absorbed into the Seagram wine and spirits empire.

In 1940 another Barton, Ronald, escaped to England ahead of the advancing Germans, again leaving the business to a Daniel Guestier. Guestier persuaded the Germans that Langoa-Barton was owned by an Irish neutral; the Germans did not harm it.

Almost as venerable an Irish clan here are the Lawtons. Abraham Lawton came to Bordeaux in 1836 from Cork, where he had been mayor. The shipping company he founded here, Tastet & Lawton, survives. Hugues Lawton is a *négociant,* or wholesaler and shipper, and his brother Daniel is a broker, or middleman, between the chateaus and the *négociants.*

In spite of their prominence in the Bordeaux wine trade, neither brother owns a chateau. Their grandfather, Édouard, owned Léoville-Poyferré and Édouard's cousin, Jean, owned Château Cantenac-Brown for many years. Their mother, Simone deLuze Lawton, is an example of how the Irish families became integrated here. The deLuzes are an old and distinguished *négociant* family.

The Phelans—the accent didn't come until later—arrived in Bordeaux around 1796 from Tipperary. Bernard Phelan bought Château de Ségur, in St.-Estephe, in 1810. His son, Frank, took over after his father's death in 1841. Frank was mayor of St.-Estephe for a time and his portrait now graces Phélan Ségur's second wine, which is called Frank Phélan. He also married a Guestier.

The chateau passed out of the Phélan family in 1918 and is now owned by Xavier Gardinier, from the Champagne region, who made much of his fortune mining phosphates in Florida.

Château Clarke, in Listrac, is a magnificent building but it has little relation to the Clarkes. Tobie Clarke was a gun merchant from County Down who came to Nantes after the Battle of the Boyne in 1690, then settled in Bordeaux. His son Luc-Tobie, a criminal court judge, gave the estate its name. The house was demolished in 1955. Baron Edmond de Rothschild bought the property in 1979 and erected the present structure.

One of the best-known Irish names in Bordeaux is Lynch. In the hands of Jean-Michel Cazes and his father, André, who bought the property in 1934, Château Lynch-Bages, in Pauillac, has become one of the most sought-after of all Bordeaux wines.

The first Lynch in France was Col. John Lynch, an authentic member of Les Oies Sauvages, the Wild Geese. He came to France after the Irish lost the battle of Aughrim to the British in 1691, turned to commerce, and did well.

The best-known Lynch was Jean-Baptiste, who was born at Château Dauzac in the Médoc in 1749. He was variously a staunch parliamentarian, a royalist, and then, when it seemed propitious, a Bonapartist. When Napoleon fell, he quickly became a royalist again. When Napoleon returned from Elba, Jean-Baptiste, who had been mayor of Bordeaux for a time, fled to England. He returned after Waterloo and died at Dauzac in 1835.

Another descendant of Jean-Baptiste Lynch, Alain Miailhe, owned Château Dauzac for a time in the 1970s. He decided to rename it Lynch-Dauzac, but was sued by the Cazes and the owners of Lynch-Moussas, the Casteja family. They won and he had to drop Lynch from the property's name. He sold it soon thereafter.

Mr. Miailhe (pronounced mee-AYE), the owner of Château Siran in Margaux, and his sister May de Lencquesaing, who owns Château Pichon Lalande in Pauillac, are also decended from the Burkes, a Galway family that made a fortune in the Philippines and one of whose daughters married a Frenchman who brought her to Bordeaux.

Asked to name a few Bordeaux chateaus with Irish connections, many people automatically mention Château Haut-Brion, in Pessac, one of the five first growths of Bordeaux. They believe Haut-Brion to be a French derivative of O'Brien. It is not. The name comes from a language spoken in the region hundreds of years ago and refers to the slight rise in the ground on which the property, once far larger, is situated.

Nor are the American owners, the Dillons, of Irish descent. Clarence Dillon, one of the founders of the Wall Street firm Dillon-Read, who bought Haut-Brion in the mid-1930s, was of Polish origin and had changed his name as a young man. After his death, the chateau passed to his son, Douglas, who was the United States ambassador to France and secretary of the treasury in the 1960s. His daughter, Joan, the Duchesse de Mouchy, directs its operation.

Thus the Dillons of Haut-Brion have no connection with the Dillons who were, with the MacCarthys, among the most prominent of all the Irish families in France in the seventeenth and eighteenth centuries. The Irish Dillons' fame came originally from their connection with the Irish Brigades. The Brigades were first organized in the mid-seventeenth century when Charles II, exiled from Cromwell's England, formed an army in exile in France. One of these was headed by James Dillon of Roscommon.

Generations of Dillons served in these private armies. Some returned to Ireland, but most of those not killed in battle eventually moved into French society. One history of the family lists seven separate Dillon families with French connections.

The MacCarthys, once said to be the richest of all Bordeaux wine merchants, on a par economically with the Hennessys in Cognac, are now known only by Château MacCarthy, a minor property in the Médoc, and by the Hotel MacCarthy, a magnificent town house in the center of Bordeaux.

One wine that will not be drunk during the March celebration is Château St. Patrick. Maurice Healey, the 19th-century Irish wine writer, claims to have found such a wine in Boston during his travels. Anthony Barton said he had never heard of it, but added, "It is a lovely thought, though."

MARCH, 1991

## FOR A WINERY, VALUE IS ALL IN THE FAMILY

When he graduated from Holy Cross College in 1973, Mike Benziger could have gone home to White Plains and built a nice life for himself. He could have joined Park-Benziger, the wine importing house founded in New York by his grandfather Joe forty years earlier and run by his father, Bruno. But no. He chose to go to California for postgraduate work in skiing and surfing. To fill in the time between the slopes and the Pacific, he worked in a wine shop, using the knowledge he had acquired from the family business. He promptly fell in love with California wine.

Within a few years, Mike had lured his entire family west to share his enthusiasm, including his parents and his six brothers and sisters. Together they founded the Glen Ellen Winery in 1981 on an eighty-five-acre parcel of vineyard land just north of the town of Sonoma, an hour's drive north of San Francisco. It turned out to be one of the more successful gambles of the last two decades in the California wine boom.

Tasting, and drinking, about a dozen of the latest Benziger releases recently, I was reminded of how innovative the Benzigers have been. Starting with a handful of premium wines made from their own grapes, they saw an opportunity that no one else in the business had discerned when a huge surplus of good but inexpensive varietal wines came on the market in the early 1980s. Under the Glen Ellen and M. J. Vallejo labels, they produced large quantities of varietal wines—wines made from specific grape varieties—selling in the $4 to $5 range. By the late 1980s, Glen Ellen, with a wine category that became known as "fighting varietals," which aggressively challenged the tradition of expensive varietal wines, was the largest winery in Sonoma County, with annual revenues of $135 million.

But this was not what the Benzigers had come west to achieve. With a production of 4 million cases a year, they were getting into the ring with heavyweights like E. Gallo, and to stay in the high-volume, low-cost game, they would need huge infusions of capital. So in 1993 they sold the Glen Ellen business to Heublein for $80 million, changed the winery name to the Benziger Family Winery, and downsized the business to today's 180,000 cases. Now they were going for quality rather than quantity.

At a time when wine prices are appallingly high, the Benzigers have managed to keep theirs within reason. Their wines represent good value. I recently tried several 1997s, including a Sonoma County merlot at $17 and a Sonoma cabernet sauvignon, $17. Both are fruity, well-structured wines that match competitors costing at least $10 more a bottle. I particularly liked a 1997 California pinot noir, $21; and also tasted a 1998 Carneros Chardonnay, $13; a 1997 Sonoma County zinfandel, $21; and a California syrah, $20.

In addition I tasted several wines of less-known varieties produced in small quantities and available only at the winery, including a white Burgundy, which I liked; a sangiovese; and a petite sirah from the "Imagery Collection." There is also a small line of reserve wines and an even smaller group of single-vineyard wines: three cabernet sauvignons and a chardonnay from California, and from Oregon, a pinot noir and a pinot gris, none of which I have tried yet.

In contrast to these small-batch wines, the standard Benziger wines usually include grapes from many vineyards. The 1997 Sonoma cabernet sauvignon, for example, includes thirty-three lots of grapes from eighteen ranches.

In California, farms are called ranches. A single grape-grower's ranch may include several well-known vineyards. Each year, Benziger buys three hundred lots of grapes from more than sixty California ranches, from Paso Robles in the south-central coastal region to Mendocino in the north. This is in addition to the grapes grown in the eighty-five-acre Sonoma Mountain

Estate home vineyard and its sixty-five-acre Willamette Valley Vineyards in Oregon.

With much fanfare, the family opened a small brewery in 1997 on the Sonoma estate to produce 100,000 cases of premium beer for the Sonoma Valley and San Francisco. It closed after a year. Mike Benziger, the winery president, said, "We opened for business just as the microbrewery fad was peaking."

The brewery is being converted into a second winery that the Benzigers say will allow them to increase production to 500,000 cases a year.

The Benzigers are proud of their reasonable prices but are not opposed to raising some of them. The top cabernet sauvignon sells for around $45. "I'd like that to go up," Mike Benziger said, "and I'd like our vineyard-designated wines to range between $35 and $70 a bottle. But we've got to earn that."

As family wineries go, not even the Gallos can claim as large an active clan. Only the Fetzers, who once had ten children involved in running their vineyards, can match the Benzigers.

Of the other Benziger brothers, Bob handles business development, and Joe and Jerry are winemakers. Of the sisters, Patsy manages a winery apprentice system, Chris is midwestern regional sales manager, and Kathy is eastern sales manager. Patsy's husband, Tim Wallace, is chief executive officer. Helen Benziger, mother of the seven, manages V.I.P. services. Bruno Benziger died in 1989.

Counting husbands, wives, and children, there are, Mike Benziger said, "more than two dozen Benzigers living in the Sonoma Valley and involved in one way or another with the winery. Our strategy is to stay family and stay small." With all those Benzigers making and selling wine, that isn't going to be easy.

MARCH, 2000

## HIS PASSIONS: FAST TRAINS AND BURGUNDY

Coming down to Burgundy from Paris on the sleek high-speed train, it occurred to me that the ninety-minute trip would not have been possible were it not for Faiveley Industries, the Paris-based company that makes doors, brakes, and other vital rail parts, not just for the T.G.V., but for trains all over the world.

When François Faiveley met me on the platform at Dijon, my thoughts abruptly switched from trains to wine. Mr. Faiveley is chairman of the industrial firm and also the director of Domaine Faiveley, possibly the largest vineyard owner in Burgundy.

Men who have made their fortune in business often reward themselves with an investment in winemaking. For the Faiveleys, the wine came first. François is the sixth generation to head the domaine since it was founded by Pierre Faiveley in 1825. In a region of tiny landholdings, the Domaine Faiveley consists of about 300 acres, some 112 of them in the Côte d'Or, the home of Burgundy's greatest wines, and the rest, more than 178 acres, in Mercurey, a few miles to the south, in the Côte Chalonnaise.

The Faiveleys have long been among the first families of Burgundy. In 1934, in the depths of the Depression, François's grandfather Georges, along with Camille Rodier, another wine shipper in Nuits-St.-Georges, founded the Confrérie des Chevaliers du Tastevin, one of the most successful promotional ventures in the modern history of wine. The Confrérie, a society with chapters all over the world, holds regular dinners to celebrate, promote, and of course drink the wines of Burgundy. Its headquarters are at the Château de Clos de Vougeot, a former monastery in the heart of the Burgundy vineyards.

Burgundy is the most complicated of all wine regions, but a closer look at Faiveley might make it slightly less so. The great vineyards of the Côte d'Or stretch from the suburbs of Dijon in the north to three little villages known as the Maranges, near Santenay, thirty-five miles to the south.

These vineyards can be traced to ancient times; some were first planted by the Gauls and the Romans. The 125 acres of grapes at the Clos de Vougeot were planted by Cistercian monks in the twelfth century.

After the French Revolution nationalized most vineyards, the laws of succession required that the new owners divide them and divide them again among their heirs. The result has been a crazy-quilt of ownership, where parcels of famous vineyards often amount to half a row of vines. From the roadway, the Clos de Vougeot vineyard seems to be a single entity. In fact, it consists of sixty-five parcels, each individually owned. One three-acre parcel is Faiveley's.

A few vineyards have managed to remain intact over the years; one of them is the Clos de la Maréchale in Nuits-St.-Georges. Faiveley doesn't own it but has a lease that runs until 2002. It accounts for about 24 of the domaine's 112 acres in the Côte d'Or (which is a contraction for Côte d'Orient, meaning the "eastern slope"). The Clos-de-Cortons–Faiveley is another single-owner vineyard. This one, owned by the Faiveleys, used to be called Rognet-et-Corton (a *rognet* is a piece of land that came to its owner as a dowry). In 1930, a court in Dijon gave the Faiveleys the right to call it Clos-de-Cortons–Faiveley.

By contrast, Faiveley's holdings in Le Musigny, a famous vineyard in Chambolle-Musigny, a couple of miles north of Nuits-St.-Georges, consist of about a hundredth of an acre and probably produce—in a good year— some twenty cases of wine.

In all, Faiveley's 112 acres in the Côte d'Or are divided among twenty-six vineyards in seven communities. In the Côte Chalonnaise, which takes its name from Chalon, a rail center east of Mercurey, Faiveley owns a number of vineyards. Two of the best known, both in Mercurey, are the fifteen-acre Clos des Myglands and the twenty-seven-acre La Framboisière. The firm owns parcels in vineyards in Givry, Rully, Montagny, and Bouzeron, all of them well-known wine-producing towns.

These communes, and particularly Mercurey, produce good and occasionally excellent wines. But their success came recently. When Burgundy was cheap, much of the wine from Rully and Givry went into sparkling Burgundy. Efforts to produce superior wines in the region, led by Faiveley, began in the early 1960s.

The great demand for fine Burgundy, which began to develop in the 1960s, brought about the revival of the vineyards on the western slopes of the hills that make up the Côte d'Or's great vineyards; they are known as the Hautes Côtes, or "high slopes." Faiveley owns about twelve acres in the Hautes-Côtes de Nuits, in the hills above Nuits-St.-Georges. Faiveley produces sixty thousand cases of wine a year, 80 percent from Faiveley grapes. Contract growers supply the rest.

Almost all Faiveley wines are red, with the best of them, including Clos de la Maréchale, Chambertin–Clos de Bèze, and Les Damodes, sold under the label Maison Joseph Faiveley. But there are a few whites, including one of Burgundy's best, a Corton-Charlemagne from vines at the top of the slope in the village of Aloxe-Corton. In a good year, there may be 150 cases of this wine.

Easier to find are the excellent, and reasonably priced, white wines from the Côte Chalonnaise vineyards. All are made from the chardonnay grape, except the aligoté from Bouzeron. Domaine Faiveley, along with Aubert de Villaine, an owner and director of the famed Domaine de la Romanée-Conti, in Vosne-Romanée, have shown that a delectable white wine can be made from this once-scorned grape.

Wandering through the cellars in Nuits-St.-Georges with François Faiveley, I came upon two men bent over the spigot of a barrel of Chambertin–Clos de Bèze. They were filling and corking bottles by hand. "We do it with all our *grand cru* wines," Mr. Faiveley said. "We won't compromise in any way with these wines." Which, of course, is part of the reason some of these wines sell for considerably more than $100 a bottle.

François Faiveley has been in charge of the family wine business since 1976, when, in a gesture that has become a family legend, his father, Guy, dropped the winery keys on his twenty-five-year-old son's desk and walked away. "I spend most of my time here taking care of the winery and the vineyards," Mr. Faiveley said. "I'm aware that I'm a bad salesman and that I've never been enthusiastic about selling the wine. My true interest is on the technical side. I make the wine myself, something I wouldn't miss for the world."

In fact, his real passion is the aroma of wine. "I think I would have been a perfume maker if I were not in the wine business," he said.

Some older hands in the trade say Faiveley wines have become richer under the direction of Francois. Perhaps, but Faiveley Burgundies made by his father and grandfather in the 1940s and 1950s turn up now and then, and they can be extraordinary.

Mr. Faiveley is also protective of his wines. In 1994 he sued Robert M. Parker Jr., the American wine critic, for implying in one of his books that there was a difference between some Faiveley wines tasted in the cellars here and the same wines purchased in the United States. The suit was settled out of court. "I had to do it," Mr. Faiveley said. "Distributors all over the world were calling to demand an explanation."

Despite that accusation, Mr. Parker was an early champion of Faiveley wines. Mr. Parker told me that as part of the settlement he had agreed not to discuss it, but, he noted, "I still rank Faiveley's wines among the best in Burgundy."

Mr. Faiveley draws on what appears to be a bottomless store of nervous energy, shuttling among the family vineyards and the wineries in Nuits-St.-Georges and in Mercurey and dashing up to Paris on Faiveley Industries matters. He expends even more energy traveling the world to promote his wines, regularly visiting the United States and the Far East.

His favorite recreation is sailing, and he crosses the Atlantic at the helm of his forty-two-foot sloop as casually as some of his neighbors drive down the *autoroute* to Nice. "The boat is in Trinidad at the moment," he said. "I'm going to move it to Tobago this summer and, God willing, sail it along the Venezuelan coast next winter." The boat is called *Glenn II,* not after the American astronaut, but after another Faiveley passion, the Canadian pianist Glenn Gould.

Wine, perfume, Glenn Gould's playing—is there room for anything else? There is.

Back on the train platform at Dijon later in the day, Mr. Faiveley patted the side of the high-speed train about to depart for Paris. "Faiveley Industries products may be less romantic," he said, "but I still remember the day the T.G.V. broke the world speed record for trains at 515.3 kilometers per hour with a Faiveley pentograph on the roof. We all had a lump in our throat."

Quickly washed down with a good Faiveley Burgundy.

JUNE, 1997

## ANOTHER ROTHSCHILD STANDS BEHIND A BORDEAUX

Who would have thought, a decade ago, that by the turn of the century two of the four Rothschilds most closely identified with Bordeaux wine would be women?

Philippine de Rothschild, the owner and director of Château Mouton-Rothschild since the death of her father, Philippe, in 1988, is the best known. But that may change. Nadine de Rothschild is beginning to make her way in the wine world. (Nadine is the owner of Château Malmaison and the wife of Baron Edmond de Rothschild, a cousin of Philippe's. Edmond is the owner of Château Clarke and a shareholder in the Domaines Barons Rothschild, which owns Château Lafite-Rothschild and other wine properties. The fourth Rothschild is Baron Eric, another cousin, who has been the managing partner in Lafite since 1977.)

Baroness Nadine de Rothschild was in New York recently promoting Malmaison, which achieved its separate identity as a chateau only in the mid-1980s. Previously it had been the name of the second wine of Château Clarke.

Château Clarke, which Baron Edmond acquired in 1973, is in the commune of Listrac—though the town line running through the property puts a portion of the vineyard in neighboring Moulis. That portion is Château Malmaison (wine chateaus traditionally have a chateau on the property, but not all do).

Wine has been made in Listrac and Moulis since Roman times, but it has never been in the same league as that made by the famous communes like Pauillac, St.-Julien, and Margaux, a few miles to the east. Mouton, which was purchased by the Rothschilds in 1853, and Lafite, which another branch of the family bought in 1868, are both in Pauillac.

In the 1855 classification of Bordeaux chateaus, Mouton and Lafite are both at the very top, among the five first growths. Neither Clarke nor Malmaison even figures in that hierarchy of the leading sixty chateaus.

Baron Edmond is said to be the richest of the Rothschilds, ten to one hundred times as wealthy as any of the others, according to *Le Monde,* the Paris newspaper. He is also a maverick, following his own path in banking and wine as well as in his private life.

His marriage to Nadine Lhôpitalier in 1963 startled French society. Traditionally Rothschilds marry women of culture and substance—women who are self-effacing and, invariably, Jewish. Miss Lhôpitalier, a middling film actress who used the stage name Tallier, came from a penniless family and had no education beyond grammar school. She had been a factory worker at fourteen, and later a model. And she was Catholic.

The story of their meeting at a Paris dinner party is legendary. His first words to her, so the story goes, were, "That's a beautiful diamond you're wearing; too bad it isn't real." Later, seated next to him, she watched him open a small pillbox. Inside the box she saw a wedding ring.

"I have no doubt that it's real," she said, "but isn't it in the wrong place?"

Baron Edmond was in the throes of a divorce at the time. Three years later, they were married. She quickly converted to Judaism.

"It would not have been possible to have the name Rothschild and be a

Catholic," she said some years afterward, "nor would it have been right for the son of a Rothschild to be half Jewish and half Catholic."

Benjamin, their only child, is now in his early forties. "He works in the bank in Paris," his mother said, referring to the Paris branch of Baron Edmond's bank, Compagnie Financière. "But he's also interested in the wine. If you're a Rothschild, the chances are you're going to be interested in one or the other—or both."

Not to be outdone by her husband and son, Baroness Nadine took a short course in enology at the University of Bordeaux before taking on the Malmaison project. She is also a successful writer. Her seven books include one best-seller in France, a memoir published in 1983 titled *The Baroness Will Return at 5.*

Buying Château Clarke and the adjoining estates, Malmaison and Peyrelebade, was typical of Edmond de Rothschild. In 1973, when the Bordeaux wine trade was in one of its periodic downturns, some famous properties were in serious financial trouble, including the chateaus Latour, Margaux, and Beychevelle. He chose Clarke instead, a virtually forgotten estate in a backwater of the wine country.

Originally known as Granges, the estate had first been planted with vines in the 12th century by Cistercian monks. In the 18th century, it was acquired by the Clarke family, which had emigrated from Ireland in 1692. They bought the property in 1750; the first chateau, built in 1810, was demolished in 1955.

Baron Edmond apparently hoped to do with Clarke in a few years what his cousins at Mouton and Lafite had taken a century to achieve. He invested a fortune in reviving the vineyards and building a magnificent winery and residence.

But neither Listrac nor Moulis is Pauillac. Writing of Clarke some years ago, the English wine expert David Peppercorn said, "The prices more closely reflect the prestige of the Rothschild name and the prodigious investment and effort that have gone into the wine's production rather than their intrinsic merit as wines."

Of the 430 acres that comprise the Clarke estate, some 360 are planted in vines. The Malmaison vineyards account for about 60 acres. The vines at both properties, abandoned for some twenty years when Baron Edmond bought the property, were totally replanted between 1974 and 1978.

Wines under the Malmaison label have appeared in the United States in limited quantities since the mid-1980s. I found good things to say about the 1988 vintage in 1993 and the 1990 vintage in 1995.

At a tasting in 1997, recent vintages, including the 1995, came off well for a *cru bourgeois,* which is what this wine is—a well-made minor Bordeaux. Moulis has long had a better reputation than Listrac for its wines. Château Chasse-Spleen and Château Poujeaux are both in Moulis. Whether Château Malmaison—so close to Listrac and once considered the second wine of a Listrac property—actually shares in Moulis's superiority, only time will tell.

For now, it is an attractive, rather one-dimensional wine at a good price. And for whatever it's worth, it is a Rothschild wine.

OCTOBER, 1997

## WITH THE ROTHSCHILDS, BIGGER *CAN* MEAN BETTER

"Mouton," said the Baroness, "was my father's *danseuse.*" Philippine de Rothschild was talking about the late Baron Philippe de Rothschild, who lavished a fortune on his wine chateau, Mouton-Rothschild. In France, a *danseuse*—a dancing girl—is a light-hearted reference to a rich man's indulgence.

Mouton-Rothschild and its renowned Bordeaux neighbor Lafite-Rothschild, owned by another branch of the family, were acquired in the nineteenth century when, for Rothschilds, money was literally no object and wine chateaus, like steam yachts, were evidence of prestige and affluence.

Those days are over. Great wine chateaus—and probably dancing girls, too—are expected to pay their way or at least to lose as little as possible. In fact, the chances of a first growth chateau ever making it on its own are slim. Costs go up continually, but the amount of wine a chateau can produce is finite: about 17,000 cases a year at Lafite-Rothschild, about 20,000 at Mouton-Rothschild, with little chance for expansion.

But there is room for acquisition of other properties and for the development of other wines, all under the watchful eyes of business managers. That is the path the Rothschilds have chosen.

Baroness Philippine assumed the leadership of Mouton-Rothschild after her father's death in 1988. Lafite has been managed by her cousin Baron Eric de Rothschild since 1977. The two cousins are members of separate branches of the family: Baroness Philippine is a descendant of the English branch, and Baron Eric, although he was born in New York, is part of the French branch of the clan. The Lafite chateau is owned by a family consortium called Domaines Barons de Rothschild.

Both properties have thrived in recent years, although thanks to her father's foresight, Baroness Philippine had a head start. While Lafite languished under caretakers, the Baron lived at Mouton and worked to put it on sound financial footing.

In the 1930s, Baron Philippe dreamed up a separate label, Mouton Cadet, for some wine he felt was not worthy of the Mouton-Rothschild name. Mouton Cadet, which means "Mouton Junior," became so popular that Baron Philippe turned it into an inexpensive Bordeaux. Current sales of about 15 million bottles a year make Mouton Cadet the best-known Bordeaux in the world, and the income it produces helps ease the burden of running Mouton-Rothschild.

In 1933 Baron Philippe added an adjoining wine estate, Château Mouton d'Armailhacq, to his holdings. He renamed it Château Mouton-Baron-Philippe, then Mouton-Baronne-Philippe after his second wife, Pauline Fairfax Potter, an American. Now the name has reverted to d'Armailhac, without the "q." In 1970 he bought another neighbor, Château Clerc-Milon, as well.

Baron Philippe de Rothschild S.A., the company he formed that now controls the family's wine interests, produces a line of regional Bordeaux, wines like St.-Émilion and Margaux. In the mid-1990s it introduced a line of inexpensive varietals—chardonnay, cabernet sauvignon, and others—produced in the south of France.

The Baron was also the driving force behind Opus One, the joint venture in the Napa Valley that he and Robert Mondavi created in 1979. (See pages 55–58.)

Baroness Philippine turns up in the Napa Valley twice a year, but she spends much of her time touring art galleries and museums around the world with an exhibition of Mouton-Rothschild labels. Beginning in 1945, her father commissioned a different prominent artist each year to design his wine label. Chagall, Cocteau, Dali, Warhol, and Motherwell are a few who contributed. (Baroness Philippine was born in Paris in 1935. Her mother, Baron

Philippe's first wife, Elizabeth de Chambure, died in the Nazi concentration camp at Ravensbruck in 1945. Baroness Philippine spent some twenty years on the Paris stage before taking on Baron Philippe de Rothschild S.A.)

Since Baron Eric took over at Lafite, he has made great strides in catching up with his friendly rivals at Mouton. In his twenty-seven years at the helm of Château Lafite-Rothschild, he has instituted a number of changes, including making better wine. Château Duhart-Milon, which adjoins Lafite, was purchased in 1962. Eventually, when the wine improved, the property was named Duhart-Milon-Rothschild.

Among other acquisitions: In 1973, Lafite-Rothschild took on a little-known chateau in the northern reaches of the Bordeaux wine country called La Cardonne. In 1982, La Cardonne was exchanged for a major share in the well-known Château L'Évangile, in Pomerol. In 1984, another famous Bordeaux property, Château Rieussec, in Sauternes, was purchased in association with a Belgian investor, giving the Domaines Barons de Rothschild a prestigious white wine estate.

In 1988, the company acquired 50 percent of Los Vascos, a five-thousand-acre property in Chile currently producing some 1.2 million bottles of wine annually. In 1987, an exchange of shares with the Chalone Group in California gave the Rothschilds a participation in four wineries there, and in 1991 they entered into a joint venture with the Quinta do Carmo, a Port maker.

Baron Eric, who travels often on Lafite business, works out of the Rothschild bank in Paris and has never been as closely identified with Lafite as Philippe was with Mouton.

Another cousin, Edmond de Rothschild, is the third wine Rothschild. He is actually part of the previous generation (he was born in 1926) and is the largest shareholder in the Domaines Barons de Rothschild. But his active involvement in wine is fairly recent. In 1973 Baron Edmond bought a tumbledown chateau called Clarke in Listrac, a commune in the Bordeaux wine country not far from Pauillac, the hometown of both Lafite and Mouton. In 1979 he added two adjoining properties, the chateaus Malmaison and Peyrelebade.

Baron Edmond, by far the wealthiest of the Rothschilds, is probably less concerned with profits than the other wine Rothschilds. A banker who lives in Geneva, he has devoted much of his life to Jewish affairs and philanthropy. A substantial portion of each year's vintage of Château Clarke is bottled as kosher wine for sale in the United States. His son, Benjamin, who was educated in the United States, helps run Clarke. Baron Edmond is also a major investor in the Savour Club, France's largest mail-order wine business. That venture is said to be successful, but it is unlikely that it can cover the investments in Clarke.

The Rothschilds were at the apex of their powers when they bought the two already famous chateaus: Mouton in 1853 and Lafite in 1868. They were mostly absentee owners of Mouton until Philippe, then just twenty, arrived in 1922. He had fallen in love with the place when he was shipped there as a boy to escape the shelling of Paris during World War I.

Poet, translator, film producer, auto racer, art collector, and world-class bon vivant, Philippe de Rothschild, with his second wife, transformed Mouton from a back-country farm to a renowned mecca for film stars, artists, royalty, and heads of state. An invitation to Mouton was prized above one to the Élysée Palace.

Two years after the Rothschilds acquired Mouton, the Bordeaux wine trade published a rating of its best wines, the famous Classification of 1855. In the first, most exclusive, group were Lafite, Latour, Margaux, and Haut-Brion. Mouton was ranked with the seconds. In 1975, after a fifty-year campaign by Baron Philippe, Mouton was upgraded to a first growth. It was the only time the classification had ever been changed. To celebrate, he reproduced a Picasso from his private collection on the label of his 1973 vintage, the wine released in 1975.

Baron Philippe may have been the last of what the French call *grands seigneurs,* wealthy men with great style. Once, in the late 1970s, he invited to lunch the owner of a nearby chateau, a threadbare elderly man who lived alone on his dilapidated estate.

As a gesture, Philippe served a bottle of this guest's wine from the rare 1945 vintage. The old man was close to tears. "We haven't had any of this for years," he said. Later, when he went to retrieve his car, he found on the front seat a case of his 1945 wine, a gift from the vast cellars of Mouton-Rothschild.

Château Lafite was known as the first among firsts in Bordeaux a century before it became Lafite-Rothschild in 1868, the year Baron James de Rothschild, a member of the French branch of the family, bought it. He paid a considerable price, nearly $20 million in current dollars, goaded, it has been said over the years, by jealousy of his British cousins at Mouton.

In Paris, Baron Eric looked back at the glory days of the family's wine chateaus. "Those were extraordinary times," he said, "but they're over. This is a business now, and the chateaus are run by businessmen."

Up to a point. Some years ago Lafite needed a new cellar for aging its precious wine. The traditional Bordeaux cellar, a large rectangular room, would have done the job. Not for Lafite-Rothschild. Baron Eric engaged the Spanish architect Ricardo Bofill to design and build—at enormous expense—the first circular wine cellar anywhere.

There is the accepted, prudent way to do things. And then there is—still—the Rothschild way.

FEBRUARY, 1996

## THE MERCHANT OF TUSCANY

Why is it that American social historians and their novelist counterparts so rarely focus on the Italians? Granted, observing a cultural milieu through the eyes of one of Edith Wharton's diffident heiresses may not be ideal, but *faute de mieux,* it's better than nothing.

Yet when it comes to the Italians and the great Florentine families, that's about what we have—nothing. And it's not that they aren't noteworthy. Many of the ancient clans of England and France are but parvenus compared with the Frescobaldis, the Antinoris, and the Ricasolis, who hobnobbed with the likes of Dante and the Medicis, and who were there at the birth of the Renaissance.

But lacking another Galsworthy or Evelyn Waugh, it's left to the lowly wine writer to limn the histories of such illustrious broods, with their palaces, abbeys, and vast estates. And truth be told, it can be a lot more fun than going on about Chianti.

The Antinoris also make the task easy. Their family tree is as remarkable as any in Florence, and their modern adventures in the wine trade are downright dizzying.

The family's current doyen, the Marchese Piero Antinori, may prove to be the most innovative of all Antinoris, who have been in the winemaking business for six centuries. With the assistance of Giacomo Tachis, an enologist and a close adviser for more than thirty-five years, Antinori has done no less than preside over a second Risorgimento—a rebirth of the wines of Tuscany.

The modern era in Tuscan wine began in the 1960s. For centuries before that, the vines of the Chianti estates had been tended by peasant sharecroppers who gave half of their profits to the owners under a system called the *mezzadria*. Each peasant's small plot was a hopeless mélange of vines, olive trees, and other crops designed for short-term gain. "It was called *cultura promiscua*," Antinori said. "It was a holdover from another time."

Inspired by Italy's economic revival after World War II and encouraged by the Communists, workers rejected the *mezzadria* and fled to jobs in the cities. It was a time of crisis—and opportunity. With the fields virtually abandoned, it was possible to consolidate them and replant.

But the opportunity, as Antinori ruefully admits, was soon squandered. The abandoned plots were indeed consolidated into large vineyards, though not with the intended results. "In seven or eight years, some 15,000 acres were planted," Antinori said, "all in sangiovese, the basic Chianti grape. But we had no experience in how or where to plant and how to train the vines. We had no controls; no one knew what to do."

And so the vintners went back to the drawing board, replanting once again, this time with the help of skilled viticulturists and winemakers like Tachis, whom the Antinoris hired in 1961. As a result, the vineyards were smaller, better located and less densely cultivated. Varieties of grapes like chardonnay and pinot noir—all new to Tuscany—were imported, while the basic techniques of Italian winemaking were being reconsidered.

The Chianti formula, for example, was an officially decreed mixture of sangiovese and canaiolo red grapes, plus 10 to 30 percent trebbiano and malvasia white grapes. The Antinoris knew that this was a recipe for mediocrity. They wanted to make intense, rich, long-lived red wines, wines that could compete with Bordeaux. "It was then that we started to experiment with smaller percentages of white grapes and untraditional aging techniques," Antinori said.

The first fruit of that experimentation was Tignanello, which appeared in 1971. When it was introduced, it was a classic Chianti blend that had been aged for a time in small French oak barrels. Gradually, white grapes were dropped, then the canaiolo. Cabernet sauvignon was added to the sangiovese, then cabernet franc, another French grape.

In 1978, the Antinoris introduced Solaia, a cabernet with about 20 per-

cent sangiovese. In 1982 came Galestro, a light white wine made from trebbiano and malvasia, the traditional Chianti grapes used in the past. Galestro
is not strictly a proprietary variety, as Tignanello is; it is made by some 15
Tuscan wine producers.

More whites followed: In 1984, a sauvignon blanc called Castello della
Sala, later renamed Borro della Sala, and in 1987, a chardonnay, Cervaro
della Sala. In 1991, the name Castello della Sala was resurrected to identify
an Orvieto Classico.

Because of their nontraditional blends, neither Tignanello nor Solaia
could be called Chianti. Further changes in Italian wine regulations in 1984
made Tignanello eligible, but the house of Antinori prefers to leave both
wines in the vino di tavola, or table wine, category.

Not that Antinori lacks for Chianti. There are three: two riservas, Villa
Antinori and Tenuta Marchese Antinori, and a single-vineyard classico
called Peppoli.

Like winemakers around the world, Antinori views the '60s and '70s as
years of technical experimentation and innovation, achievements that were
then exploited in the '80s. Now, he says, "It's time to return to the vineyard, the real key to quality winemaking."

At 55, Antinori has many years of winemaking ahead of him. And
when the social historians of the future get around to Florence, they'll discover that the Antinoris' seventh century was at least as lively as their first six.

NOVEMBER, 1993

## A PRINCE OF CHAMPAGNE

Winemakers come in many shapes, sizes, sexes, and political persua-
sions, and I thought I'd pretty much come across them all. Until
someone mentioned that the chief enologist for Pommery Champagne was
a prince.

An authentic prince?

I was skeptical. As a result of mergers upon mergers in recent years, most
of the old-line Champagne houses are run by, well, bean counters. Still,
some of them are clever enough to keep a title or two around to convey the
image of privileged elegance that Champagne has always striven for.

Would Alain de Polignac turn out to be just another titled frontman, tanned from the Côte d'Azur and eager to gossip about polo and the price of a new Aston-Martin? I called from Paris and Prince Alain, as everyone at Pommery calls him, from the forklift drivers up, agreed to spend some time talking about Champagne and, I hoped, the Polignacs.

"Good morning, good morning," he called, striding across the winery's park. "Thank you so much for coming to see us." Sixtyish, trim, and with a striking resemblance to Noel Coward, Prince Alain was the model of a successful businessman, down to what had to be Savile Row tailoring. Around his neck and tossed insouciantly over his left shoulder was a scarf of vivid yellow cashmere—his trademark, I later learned.

"Isn't it lovely here?" he asked as we walked through the gardens to his office. It was in fact a lovely early spring morning. In the distance was Rheims, dominated by the immense cathedral where so many kings of France had been crowned. Just before us were the buildings of Pommery, looking much as they must have when Mme. Louis Pommery had them built more than a century ago. Prince Alain laughed when I voiced my egalitarian suspicions about him. "I am a chemical engineer," he said. "I am also an enologist and have been the chief enologist at Pommery for ten years. I also happen to be a prince."

Pommery was founded in 1836 by Alexandre-Louis Pommery, a Rheims textile merchant, and Narcisse Greno, a wine merchant who eventually left the business. When Pommery died in 1858, his widow, then thirty-nine, took charge. Over thirty years, she built a modest little business specializing in red wines into one of the largest and most prosperous of all Champagne houses. Today, it produces some seven million bottles a year.

The Polignacs, who trace their history back to 809, came to Pommery in 1879 when Mme. Pommery's daughter, also Louise, married Count Guy de Polignac, later a marquis. A son of that marriage, Melchior, also a marquis, ran the company from 1907 to 1947. In 1952, one of his nephews, another Guy de Polignac, took over and ran Pommery until 1979. Since 1991, Pommery has been owned by LVMH (Louis Vuitton Moët Hennessy), which also owns Moët & Chandon, Dom Ruinart, Veuve Cliquot, Krug, and other Champagne producers. The second Guy de Polignac ran Pommery with his two brothers, Louis and Édmond, all of them princes. Alain is Édmond's son.

"My title came to us from the king of Bavaria," he said. As if that would clear things up.

In 1972, when Patrick Forbes wrote about Pommery and the Polignacs in *Champagne,* he noted that the family seat was the Château de Polignac in the Loire Valley but that there were "no less than fifteen Polignac ancestral homes scattered about France." Prince Alain is proud of his family background but proud, too, of his winemaking heritage at Pommery. "In almost two centuries, there have been only eight winemakers," he said. "Each of us has held the job for twenty to thirty years; each of us has spent his last ten years training his successor." He must have about five years to go because he has been working for the last five years grooming Thierry Gasca, now the Pommery cellarmaster.

"There is no father-to-son line, here," Prince Alain said. "The only link is taste. It takes years to understand a house style—in this case, what is Pommery and, more important, what is not.

"Actually, there are two lines at Pommery, the family line and the taste line. In me, for the first time, the two are joined." And what is Pommery? "Ah, a light Champagne, yes, but full-bodied. And fresh, always fresh and lively on the palate." Over lunch we sampled "Louise," Pommery's tête de cuvée, or top-of-the-line bottling, and the vintage brut. Mme. Pommery was the first to introduce the brut, or very dry–style Champagne in the 1870s. It became the basic style of almost all the Champagne houses.

The Pommery winery covers some two hundred acres of what was once a city dump on the outskirts of Rheims. The striking buildings, erected over a ten-year period in the 1870s, sit atop a network of immense chalk caves from the Gallo-Roman era, two thousand years ago. Some twenty million bottles of Champagne are stored in them now.

Often derided by modern critics, the winery buildings are in a neo-Elizabethan style popular in England in the nineteenth century. They were meant to symbolize Pommery's strong ties to the English Champagne market. "They give a face to Pommery," Alain de Polignac said. "The style is in the Champagne. Architecture connects the look of Pommery and the style. "I would have loved to be an architect," he said, adding quickly: "Of course, I am an architect; each year I build my wines." Prince Alain may look for all the world like a prosperous businessman, but his Champagnes and, of course, his yellow scarf, say he is a sensualist. He agrees.

As we walked out of the winery, he paused. "Listen," he said, dropping his voice dramatically. "Hear that? Dring, dring, dring—the bottles bumping along the bottling line? For me, each year, it means the beginning of spring. Then in the fall, it's the smell of the new wine. Sight, sound, smell—all come together right here. Ah, I adore it."

MARCH, 2001

# CHAPTER 4

# THE SOMETIMES NASTY, SOMETIMES DELIGHTFUL, BUT ALWAYS INTERESTING BUSINESS OF WINE

Wine, it's pretty safe to say, is a business like no other. Could the law, or dentistry, or the military produce the late Walter Taylor, the redoubtable Baron of Bully Hill? Or a strange phenomenon like nouveau Beaujolais? Or Parisian thieves who steal only First Growth Bordeaux and Grand Cru Burgundies? Or a Bordeaux chateau owner like Jean-Michel Cazes, who once studied petroleum engineering in Texas?

And what about Mark Miller, a onetime magazine illustrator from Oklahoma who fell in love with wine in France and came home to build a winery, Benmarl, and restore one of the oldest vineyards in the United States less than a hundred miles up the Hudson River from New York.

Then there is hard-driving, litigious Jess Jackson, who worked as a policeman while studying law and got into wine by knowing how and where to acquire land. And who, now, thanks to his hugely successful Kendall-Jackson and a host of smaller luxury wineries, is one of the richest men in the country. Then, too, there is the Napa Wine

Company, in the heart of the Napa Valley. A winery for winemakers who don't happen to own a winery. Dozens of the Valley's most famous wines are made in this anonymous but highly efficient wine factory. And there is Beaune, the true wine capital of Burgundy, which has to be France's most popular one-industry town. Only Rheims and Épernay, both in Champagne, probably have more wine hidden in ancient cellars underneath the busy streets.

## WHAT, NO PROFITS?
## LET'S PARTY ANYWAY

The financial performance of the Chalone Wine Group, a publicly owned company, had not exactly been remarkable in recent years. So there was some concern among the 1,200 stockholders who had gathered in May, 1994, when William Hamilton, the chief financial officer, rose to speak.

"Oysters," Mr. Hamilton said, "remain unchanged, at 2,500, despite a 25 percent decline in shuckers."

The stockholders cheered and applauded.

"Salmon is up 3 percent, to 650 pounds, and leg of lamb derivatives are down 3 percent," Mr. Hamilton continued, to even more enthusiastic applause.

That's the way it is at Chalone annual meetings. Monetary matters, no matter how gloomy, are brushed aside and frivolity reigns. The meeting is held under a huge tent, and the numbers Mr. Hamilton quoted alluded to the dishes to be served.

The Chalone Wine Group comprises Chalone Vineyard in Monterey County, Acacia Winery in Napa, Carmenet Winery in Sonoma, and part ownership of Edna Valley Vineyard in San Luis Obispo, all in California; as well as a 50 percent interest in Canoe Ridge Vineyard near Walla Walla, Washington, and a joint investment with Domaines Barons de Rothschild, owners of Chateau Lafite-Rothschild and other wine properties in France, Portugal, and Chile.

Chalone went public in 1984 and enjoyed some barely profitable years before going into the red in 1992. It remained there in 1993: The company reported $18.3 million in sales but had a loss of almost $700,000. The stock, which traded as high as $11 in 1991, dropped to $5.25 in the fall of 1993 and closed at $5.50 in over-the-counter trading the day before the meeting.

"It's not exactly at an all-time high," said Richard H. Graff, Chalone's chairman, evoking yet another round of laughter. He was warmly applauded when he predicted, vaguely, that Chalone would again make some money "in a couple or three or four years."

Chalone has never paid a dividend and is unlikely to do so for many years. Even so, the stockholders are as loyal as Red Sox fans.

And why not? They are courted and coddled like matrons at a pricey spa. For openers, anyone owning at least one hundred Chalone shares is entitled to discounts of 25 percent or more on wines from any of the wineries in the group, including Château Lafite-Rothschild and its European partners.

Stockholders regularly are offered special bottlings of wines made just for them. And they receive dividends with each wine purchase that can be applied to rare wines offered each fall from the Chalone library.

They can take special tours of any of Chalone's wineries, including those in France, Chile, and Portugal. The tour of Quinta do Carmo, the Rothschild property in Portugal, sold out in one day in 1993.

Some tours venture beyond Chalone affiliates. One shareholders' group toured Australia and New Zealand last year. Soon, there will be a Mexican trip: Chalone recently entered a marketing agreement with Monte Xanic near Ensenada, Mexico, a winery that is owned by a major Chalone stockholder.

Then there are the shareholder dinners held in various cities throughout the year; and shareholders can use the various Chalone wineries for parties of their own.

But the biggest event is the annual gathering at the Chalone Vineyard. The first was held in 1969; six people showed up and Mr. Graff, the chairman, cooked lunch. The 1994 attendance of 1,200 stockholders was down from 1,350 in 1993, but the price was higher: $65 a person, up from $50.

The Chalone vineyard and winery are in a remote section of the Gavilan Mountains, about three hours' drive south of San Francisco. There is a paved road up from the Salinas Valley floor, but most stockholders elect to spend the weekend in Monterey and use the air-conditioned buses provided by the winery. A few fly in, using the Chalone airstrip, but winery officials frown on this; the landing is not an easy one.

After guests stake out a table in the main tent, they spend several hours wandering around the property, from one smaller tent to another, sampling the dozens of wines made by members of the Chalone group. They meet old friends, make jokes about the company's financial situation, and—very important—they buy wine.

"We figure about $300,000 worth today," said Philip Woodward, the president of Chalone Wine Group and the man behind the unusual stockholder program, at the 1994 gathering.

Making wine is not especially difficult; distributing it and selling it, particularly for smaller wineries—or a group of small wineries—is a formidable task. The Chalone stockholders are not only loyal customers but also an enthusiastic sales and promotion staff. Chalone executives are careful to label the annual stockholders' gathering a "celebration" lest they agitate their bankers and the Securities and Exchange Commission. There is, in fact, a traditional annual meeting to deal with the hard facts and, at the moment, unpleasant realities. It was held two days prior to the party, but only thirty or so shareholders showed up.

"We have ten thousand stockholders now," Mr. Woodward said, "up about one thousand just since last year." Every spring, the number of new stockholders surges, he continued, adding, "They call and say, 'Am I too late to get invited to the meeting if I buy my shares now?'"

Dick Graff said: "We tried holding the meeting at several of the other wineries, but so many people came, things got unmanageable. We came back up here in the mountains because it's so remote, but it didn't help. The crowd just got bigger."

In his remarks to the shareholders, Mr. Woodward noted that they came from thirty-three states and five foreign countries. "There are 1,200 of you here," he said. "I thought you'd like to know that General Motors had 750 at their meeting, and I.B.M. 1,000. AT&T was tied with us at 1,200. The only company to beat us was Ben & Jerry's, the ice-cream people in Vermont. They had 2,000, but I'll bet that 1,800 of them were under ten years of age."

MARCH, 1994

## WHERE PROFITS STAY IN THE
## BOTTLE FOR 17 YEARS

A few of the original investors in Sonoma-Cutrer, the prominent California winery, probably wish they had taken their profits in old croquet balls. Or had consumed their money's worth. In the first seventeen years that it made wine, Sonoma-Cutrer produced some good and extremely popular wines. And it was the host of spectacular croquet matches played on its two international-class croquet lawns. But profits? There really weren't any.

"At our 25th reunion, my class at Harvard Business School gave me the Long Suffering Award for the seventeen years it took me to break even," said Brice Cutrer Jones, the winery's founder and chief executive.

Since Brown-Forman, a big Kentucky wine and spirits company, bought Sonoma-Cutrer in April 1999, Mr. Jones's financial tightrope act is probably over. But over a glass of his wine at Jean-Georges in New York last week, he insisted that he plans to keep doing what he has done from the start: pour almost all the profits back into the winery.

Sonoma-Cutrer Vineyards started out in 1972 as a tax shelter affair. Mr. Jones, with his new M.B.A. degree, worked in Sonoma County gathering up vineyards and land with vineyard potential, while a financial partner in New York lined up investors. Soon, almost two hundred acres of northern Sonoma were under vines.

Like many grape growers, Mr. Jones decided that the only way to make the wines he wanted—and some profit—was to make the wine himself. In 1981 Sonoma-Cutrer Vineyards made its debut in Windsor, eighty miles north of San Francisco.

Around that time, the federal courts ended the great tax shelter schemes of the 1970s. With some effort, new investors were found. Fortunately Sonoma-Cutrer wines were a hit almost from the start, largely because of Mr. Jones's choice of William Bonetti as his first winemaker. The story is told of Mr. Jones going through stacks of résumés, throwing most out unopened, until he came to the one he had been hoping for, from Mr. Bonetti. Mr. Bonetti had gained a reputation for great wines at Charles Krug and later at Souverain, and he shared Mr. Jones's almost obsessive insistence on perfection. He particularly liked Mr. Jones's radical idea of concentrating on a single wine, chardonnay.

Brice Jones, a former Air Force fighter pilot, brought other assets to the table: faith in marketing and promotion, and the blend of determination and boyish charm that an earlier generation used to say could sell refrigerators to Eskimos. Even before he had wine to pour, Mr. Jones visited wine writers and restaurants, telling them his was going to be the best wine ever. And he delivered.

Sonoma-Cutrer chardonnay began a rapid climb to the top of the popularity charts, and there it sits, year after year. *Wine & Spirits* magazine reports that it is the most frequently ordered wine in well-known American restaurants. Mr. Bonetti's protégé, Terry Adams, now makes it.

In his *Wine Atlas of California,* James Halliday, the Australian winemaker and writer, describes the attention to detail at Sonoma-Cutrer as "fanatical," with the "unbelievably rigorous selection" of grapes, the tunnel through which they pass to the presses, the winery's almost hospital-worthy sterility, and the special machine that dusts each cork before it is pressed into the bottle.

"One could write an entire book about Sonoma-Cutrer, so extraordinary—and at times bizarre—is the whole venture," Mr. Halliday said. He was also taken with the winery's croquet lawns, where the American Croquet Association's annual open raises substantial sums for charity and garners publicity for the winery.

Sonoma-Cutrer produces seventy thousand cases of wine each year, all of them from its own grapes. Some fifty thousand cases carry the name Russian River Ranches and are made mostly from grapes from three of Sonoma-Cutrer's five vineyards: Vine Hill, Kent, and Shiloh, all in the Russian River Valley viticultural area. The winery's most famous wines are Cutrer and Les Pierres, from vineyards of the same names. Cutrer is in the Russian River Valley, and Les Pierres in the Sonoma Valley; ten thousand to eleven thousand cases of each are made each year.

If anything has informed Mr. Jones's winemaking philosophy, aside from the skills of Mr. Bonetti and Mr. Adams, it is the exposure he gets from meetings with Burgundian winemakers. Every two years since 1986, Mr. Jones has sponsored chardonnay symposiums that alternate between Sonoma-Cutrer and Burgundy wineries. And the French growers have had an effect on the style of Cutrer and Les Pierres.

To illustrate the French influence on his thinking, Mr. Jones eagerly opens bottles of Corton-Charlemagne and Bâtard-Montrachet from Burgundy producers. His goal is to emulate the Burgundian ability to make white wines that combine intensity with leanness and elegance. Even he agrees that California chardonnays are often too heavy-bodied to be elegant. But some critics have said that starting in the late '80s, Sonoma-Cutrer wines have been leaner at the cost of truly intense flavors. At lunch last week, however, there was no need to open French wines. The Sonoma-Cutrer wines, the Les Pierres in particular, were "Corton-y," as Mr. Jones had promised.

(In fairness to Sonoma-Cutrer and other top California chardonnays, it should be noted that the French don't always achieve their goals. White wines from the northern half of the Côte d'Or, like Corton-Charlemagne and the rare Musigny blanc, are often more dependable than those from south of Beaune. Wines from Meursault and the Montrachet communes can occasionally be insipid.)

While Mr. Jones was reluctant to be specific, it would appear that his French connection may soon bear fruit in new ways. Most Burgundian white wine producers also make red wines. Many are even better known for their reds. Sonoma-Cutrer has prime land set aside in the Sonoma Coast appellation, and it would be no surprise to see Sonoma-Cutrer end its long monogamous affair with the chardonnay grape and take up with the seductive pinot noir, Burgundy's only red wine grape.

NOVEMBER, 1999

## THE NOUVEAU MANIA

It took me a dozen years to catch on. Every November 15, the Beaujolais nouveau would be released. Everyone knew about the midnight frenzy: the bottling lines still rattling, the frantic loading of the trucks under the glare of the television lights, and then the careering journey off through the night to the bars of Paris, to the planes at De Gaulle and the ships waiting impatiently at the Channel ports. All over the world, thirsty fans were waiting.

Then, as dawn broke over Paris, the sign would go up in the window of my Nicolas shop on the Rue St. Antoine: "*Le Nouveau Beaujolais est arrivé,*" and there, inside, row by row, were the cases of the new wine. "Remarkable," I would think. "How do they get it from Villefranche and Romanèche-Thorins to the Fourth Arrondissement so fast?"

Later, in New York, I was gulled once again. Here were cases of this new wine, fresh from the vats in the French countryside, all over Manhattan within hours of the release time in France. "We use the Concorde," one importer would say. "Chartered jet," bragged another. "Chartered jet, helicopter to midtown, and a couple of marathon runners right to the front door," announced a third with a straight face.

Eventually I learned the embarrassing truth: The November 15 deadline was mostly a myth. Some of the wine is shipped at least a week in advance. The cases are stacked in warehouses in Paris, London, even Singapore, long before the hokey race through the night begins. The inviolable release date is basically a fantasy, meant to build up excitement and lure the naive press.

Once, the deadline probably meant something. After all, the wine deserves a little consideration. But today, commercial pressures on the winemakers are too great. If the release date is November 15, or 20, or whatever, that's the day the customers want it. Beaujolais producers insist that they follow the rules. "We ship it," they acknowledge, "but we don't sell it to the public until the release date." In fact, the release date is now, and probably will remain, the next to last Thursday in November. Nature has nothing to do with it; this is a marketing decision. It makes the following weekend something of a Beaujolais nouveau weekend.

But what about the wine, the subject of all this fuss? Is there anything to this Beaujolais nouveau?

Well, it depends. As a gimmick, as a way to lighten up a gray November day, as a topic of conversation, it's first-rate. As a wine, it usually isn't very much of anything. It's never really bad, and it's never really good. It's pleasant.

Beaujolais nouveau is an unfinished wine. Both the traditional Beaujolais and the wine destined to be *nouveau*—which means "new"—are fermented at the same time, from the same gamay grapes, right after the harvest in September. The traditional wine is left to mature, usually in glass-lined concrete tanks or in stainless steel, through the winter. It's released in early spring. By then, it's a full-fledged wine. It will remain fresh and lively for a year, and if well maintained, may be good for at least a couple of years. The nouveau, at its best, will last little more than six months. It can be enjoyed for its freshness, but even that is qualified. Originally the nouveau was wine the growers and shippers did not consider worth bottling. So they sold it, still in the barrel, to the bistros in Lyons and, later, in Paris. Bottled nouveau, young as it is, conveys only a hint of the lively taste of that never-bottled young wine.

The people who benefit most from the nouveau craze are the Beaujolais producers. In the past, they lived like many farmers: on credit. They borrowed money to grow their grapes and make their wine, and they paid off their debts the following year, when they sold the wine. Today they sell fully half of their wine within two or three months after they make it. They are the only winemakers in the world who can pay cash for Christmas presents.

Inevitably this cuts down on the amount of wine permitted to mature into true Beaujolais. However, a lot of what is sold as nouveau actually comes from the southern part of the Beaujolais, a region not known for the highest-quality wine.

The nouveau phenomenon has little direct effect on the best Beaujolais, the nine *crus,* or village name, wines. Any producer who is entitled to call his wine Chénas or Moulin-à-Vent or Fleurie is unlikely to want to sell it off as nouveau. If he did, he would have to give up the prized place name; none of the nine *cru* names can be used in conjunction with *nouveau.* There is no such thing as Chénas nouveau or Fleurie nouveau.

However, the nouveau may indirectly affect the *crus.* By virtually halving the amount of wine available for traditional Beaujolais, the nouveau may well be responsible for increasing prices for all other Beaujolais. (The other *crus,* by the way, are Juliénas, Saint-Amour, Chiroubles, Morgon, Brouilly, and Côte de Brouilly.)

Drink the nouveau, by all means. Drink it in restaurants and bars, and bring a few bottles home. After that, forget it and go out to buy some real Beaujolais.

NOVEMBER, 1986

## THE BARON OF BULLY HILL

Dinner was supposed to be at eight. At nine and again at ten, the guest of honor called, hopelessly lost in the New Jersey suburbs. We talked him in, so to speak, and close to eleven he appeared, in boots and cowboy hat, swearing mightily and with a case of wine under each arm. He was Walter S. Taylor, and he had arrived.

Walter, the self-styled Baron of Bully Hill, his winery in the Finger Lakes region of New York, had invited himself. It would be the perfect way to show off his wines, he had said. He ended up presiding over a cold dinner and a tasting that, perfect or not, lasted until almost five A.M. Then he raced off for, he said, another engagement.

That was years ago. Walter Taylor—outrageous, flamboyant, single-minded, immature, irritating, charming, died in April 2001 at his home in Hammondsport, New York, at the age of sixty-nine. He had been a quadriplegic since a 1990 minivan accident in Florida. He was the scion of an old upstate New York family. His great-grandfather had been a cooper. His grandfather, the first Walter Taylor, founded the Taylor Wine Company in 1880 and built a business making wine from native labrusca grapes like the Concord. Labrusca can withstand harsh eastern winters but produce a harsh-tasting wine. To make labrusca wines more palatable, they are often cut with water, California wine, or grape juice concentrate.

In the 1950s, most winemakers believed that delicate European vines like chardonnay and merlot could not survive the bitter winters of the Finger Lakes. But a few farmers were experimenting with French-American hybrids, crosses between rugged American varieties and the European vines. Walter believed that the wines from these hybrids were vastly superior to what his family and the rest of the New York State wine industry had been making for seventy years, and he began saying so, loudly and often, to anyone who would listen. In 1970, fed up with his carping, the family banished him from the Taylor Wine Company. Seven years later they sold the business to the Coca-Cola Company. Meanwhile, Walter and his father, Greyton, bought back the first Walter Taylor's original vineyard, Bully Hill, just outside Hammondsport, on the high hills overlooking Keuka Lake. They planted it in hybrids. From his hybrid grapes, Walter produced an array of wines with names like Old Trawler White, Meat Market Red, Space Shuttle Rose, and Le Goat Blush. These names, along with the labels he designed for them and his unflagging self-promotion, made Bully Hill the best-known winery in New York State. But he continued his attacks on the New York State wine industry, particularly the Taylor Wine Company. At one point, he had a railroad tank car hauled up the hill to his vineyard to show visitors how other local winemakers, rather than use hybrid grapes, shipped in California wine to blend with their product. After buying Taylor, the Coca-Cola Company wasted no time before going after Walter. In short order, a federal court judge enjoined him from just about everything but wearing his farmer's overalls. Most serious of all: he could no longer use the name Taylor on any of his wines. Undaunted, he gathered two hundred fans and, at a highly publicized tasting, had them ink his name off hundreds of bottles of Bully Hill wine. He designed a label bearing the picture of a goat and the slogan: "They Have My Name and My Heritage, but They Didn't Get My Goat."

*Decanter,* a British wine magazine, declared it to be the ugliest label of the year. Other labels depicted him as The Lone Ranger, with the inevitable caption: "Who Was That Masked Man?" He distributed bumper stickers that said, "Enjoy Bully Hill, the Un-Taylor." Another label showed an owl and was inscribed "Walter S. Who?" His levity soon earned him a contempt citation from the court, and he was ordered to turn over to the Taylor Wine Company all the offending material he had produced. "We brought it down to them in a manure spreader," he said at the time.

"I have been thrown out of the New York State wine industry," he declaimed to me, "out of my local club, even the Hammondsport Episcopal Church. Everybody in my family, except my father, my ten-year-old son and my winemaker deserted me. The community completely rejected me and treated me like a dog. Why? Because I wanted honesty and integrity in the wine business." It was a well-rehearsed speech, one he clearly relished. It was also irrelevant because Walter was only a transitional figure in the industry. He was right about labrusca grapes; they had had their day. But the hybrids he championed so lustily were transitional, as well. Hermann Wiemer, whose family had made wine in Germany for generations, joined Bully Hill as winemaker in 1968, producing only hybrid wines. But having grown up in Germany's far northern vineyards, he rejected the idea that vinifera grapes could not survive in the Finger Lakes climate.

Besides, he knew that another winemaker, Konstantin Frank, had successfully grown vinifera near Hammondsport in the 1950s. In 1973, while continuing to work at Bully Hill, Mr. Wiemer started a small winery of his own on nearby Seneca Lake, making vinifera wines, especially riesling. On Christmas Eve 1979, while Mr. Wiemer was with his family in Germany, Walter fired him, presumably for his disloyalty to hybrids. Bully Hill offers some vinifera wines now, but it is still firmly rooted in the hybrid culture.

And for Bully Hill, it still works. Under Walter's widow, Lillian Taylor, the winery turns out around two hundred thousand cases of wine and, in its restaurant and museum, entertains some one hundred and fifty thousand visitors a year. "On weekends you can't get into the parking lot," a friend told me. The Baron may be gone, but Bully Hill is booming. Not a bad legacy for a gadfly.

MAY, 2001

## THE "DO IT YOURSELF" WINERY

Route 29 is the Yellow Brick Road of California wine. Between the city of Napa and the town of Calistoga, about twenty miles to the north, Route 29 offers a display of some of the world's most beautiful and prosperous wineries. From May to October, the road is packed with tourists come to gape at these temples of the good life, with their manicured vineyards and welcoming tasting rooms.

Hardly any of the tourists bother to glance at the Napa Wine Company as they drive by. It's not hard to understand why. Compared with the sleek Opus One and Robert Mondavi wineries just to the north, the Napa Wine Company looks like a truck terminal. But in fact, many of the most sought-after wines—the wonderful little-known labels sold through exclusive mailing lists—are made right here behind these unprepossessing walls.

Pahlmeyer Winery, the Lamborn Family Wine Company, Fife Vineyards, La Sirena, the Staglin Family Vineyard, Larkmead Vineyards, Mason Cellars, the Bryant Family Winery, Anthology, Azalea Springs, Buehler Vineyards, Broman, and Stone Creek are just a few of the fifty or so wines made in this industrial-looking space. Larger wineries like Domaine Chandon, Benziger, and Mumm Napa Valley also use the Napa Wine Company from time to time for part of their winemaking.

The Napa Wine Company is what is known in the trade as a custom-crush operation. "Custom crush" may sound like a hug from Jesse Ventura, but it means making wine in someone else's winery. For a small winemaker, the economic benefits of this are obvious. The capital investment in winemaking equipment—crushers, fermenters, bottling lines, and such—is considerable: hundreds of thousands of dollars, whether the goal is to make one hundred cases or one hundred thousand.

For a fee, the Napa Wine Company can handle the entire winemaking process from the time the grapes arrive at the winery door until, two or three years later, the bottled, aged, and labeled wine heads off to customers or wholesalers.

Charges are based on the number of tons of grapes crushed, the number of gallons of wines stored, and the number of bottles filled and packaged. Special crushing methods will cost more. "Prices can vary a lot," said Robert Lawson, the general manager, "but it usually averages out to about $24 a case."

Jason Pahlmeyer, of the Pahlmeyer Winery in St. Helena, California, is a customer and avid fan of the Napa Wine Company. "A lot of my colleagues are not eager to have it known that they use Napa Wine," he said. "They want to keep that image of the little stone winery with the ivy and bougainvillea all around, and let's face it, Napa Wine is not the prettiest place around."

But that does not mean that the company is not modern. "They've got a press over there that cost about $1 million to install and $60,000 a year to maintain," Mr. Pahlmeyer said, "and they have a state-of-the-art Italian bottling line I could never match—nor would I want to. There's no way I could afford equipment like that for my little operation, and I have one of the bigger boutique wineries." Mr. Pahlmeyer makes about five thousand cases of wine a year, but his only long-term capital investments are his barrels and a gadget he inserts in the Napa Wine bottling line to position his labels properly.

Because of an unusually cool summer, the 1999 harvest was about three weeks late, but Napa Wine was busy scheduling the arrival of grapes for dozens of growers.

"All we ask is twenty-four hours' notice," said Mr. Lawson, the company's general manager. "Traffic can get heavy when the grapes start coming in, and the growers tend to get nervous. We tell them to be patient, that everyone will be taken care of."

Napa Wine has three winemakers, including Mr. Lawson, plus several dozen assistants, but each client must have its own consulting winemaker. The consulting winemakers, who have included such stars as Tony Soter, Heidi Barrett, Helen Turley, and Cathy Corison, may take over most of the winemaking tasks or merely direct them. They also return regularly to check on their wines as they mature in barrels.

For the most part, Napa Wine employees said, their relations with the consulting winemakers are good, but some, not to be named, can be arrogant and, they say, unreasonable. "Most of them hope one day to get their own wineries," one Napa Wine staff member said. "Some, we're glad when they do."

The Napa Wine staff, most of whom have been working in Napa Valley wineries all their lives, plot their winemaking schedule with military precision. "We know what varieties will mature first and in what parts of the region," Mr. Lawson said. "Sauvignon blanc comes in first, then Napa chardonnay, then up-valley zinfandel. Pinot noir should be next, then Sonoma chardonnay, followed by Carneros chardonnay and Sonoma Coast pinot noir. Finally, we get cabernet sauvignon, merlot, and cabernet franc."

By federal law, wine for commercial use must be made in a bonded—that is, licensed—winery, and each of the small wineries that use the place have their own bond number assigned by the federal government. At harvest time, many wineries can be making wine at the same time in a twenty-four-hour-a-day operation.

When one of them is making wine in the plant, workers paint a yellow line on the floor, and that becomes that particular winery's bonded quarters for as long as the job takes, even if it's only for an hour or two. When the wine is moved into barrels for aging, the storage area then becomes the bonded premises.

"This is a very popular way to make wine," said Stephanie Grubbs, Napa Wine's sales and marketing director. "We usually have a waiting list of ten to fifteen prospective clients."

The Napa Wine Company is not alone in the custom-crush business. There are competitors in both the Napa Valley and nearby Sonoma. In addition, many larger wineries rent out their services.

Napa Wine is owned by the Pelissa family, farmers and grape growers in the Napa Valley for four generations. The present winery was built in 1870 and had been owned by Heublein & Company, which used it to make chardonnay for its Inglenook brand. The Pelissas, who own some 650 acres of grapes in the Napa Valley, bought it in 1993 and turned it into a custom-crush business.

In 1999, catching on to a trend of their own making, the Pelissas began to market their own wines under the Napa Wine Company label—a 1997 sauvignon blanc, a 1997 pinot blanc, and a 1996 cabernet sauvignon. With an eye on making top-quality wine, they are now competing with their own customers, all to the benefit of the wine drinker.

AUGUST, 1999

## CAVE DWELLING

A recent news item told of a California priest, now deceased, who lived simply and frugally—according to the dictates of his vows—but who kept an air-conditioned home for his dogs.

There are wine collectors just like the good father; they take better care of their bottles than they do of themselves. Well, almost. A prominent lawyer in Coral Gables, Florida, feuded with the city fathers, who refused to allow him to expand his home. He didn't want the extra space for him-self—it was for his wine. A New Orleans businessman moved out of his beautifully restored house in the French Quarter into an even more beau-tifully restored house three blocks away. The old home? It's now his wine cellar. A San Francisco dentist employed for many years a full-time assistant. Dental assistant? No, a cellar-man to care for his vast collection of wines.

Storing wine has become a serious adjunct to buying and drinking. A minor industry has sprung up in recent years, eager to supply wine racks, wine refrigerators, special air-conditioning, lighting, and even turnkey wine cellars ready to receive those cases of Lafite and Pétrus that have been sharing space with the galoshes in a hall closet.

The lawyer, the businessman, and the dentist are special cases. They are world-class collectors. Many of their wines are great rarities, old fragile wines that, like old fragile people, need special care. But many far less seri-ous wine buffs have elaborate storage facilities for their bottles, and as more and more occasional drinkers become collectors, new and ingenious ways to house all those bottles must be found. For suburban collectors and city folk with country homes, the problem is one of expense: Should the cellar be utilitarian (concrete walls and rough wooden racks) or elegant (a back-lighted showcase for the rare and not-so-rare bottles)?

Make no mistake: Good wine needs little more than constant tempera-ture and an absence of sunlight and vibration to last a long time. Redwood racks, special ventilation, indirect lighting, and old Burgundy posters on the walls may do wonders for the ambiance but they do very little for the wine, unless there are special problems.

Another Florida collector has special problems. He lives on the top floor of one of those high-rise Miami Beach condominiums. His wine cellar is literally on the roof of a building, where the average temperature is close to that of Abu Dhabi in the hot season. Thick insulation and enough air-conditioning to freeze the Astrodome keep his wine in good shape—heaven knows at what cost. Even so, he has it better than most apartment dwellers who are literally limited to the infamous hall closet or the space under the sink.

Some old prewar apartment buildings, particularly in New York City, provide secure basement storage space but don't guarantee the temperature. A journalist with an impressive wine collection from his foreign-correspondent days thought himself lucky to find an old West Side apartment building with safe basement space. During one particularly cold winter, a window broke near his stored wines; he lost dozens of irreplaceable bottles when the wine froze.

Parisians have a more serious problem. There, as might be expected, many apartment buildings provide special facilities for the tenants' wine. But there are gangs of thieves in Paris who specialize in raiding these cellars. Sometimes with inside help, they break in at night, on weekends, or during vacation periods and are gone within minutes. What's more, like skilled jewel thieves, they know their métier. They ignore the Macon and carry off the Meursault; they grab the great 1982 Bordeaux and leave the indifferent 1984s. As one American couple living in Paris discovered—after they'd been robbed—they were practically the only people in the building who used their *cave*.

Storing wines, in Paris, New York, or anywhere else, is less a problem than it seems. It makes sense to keep some wine around—it's bulky stuff to cart home bottle by bottle—and you never know when you will need one more bottle than you have on hand.

It's also often cheaper by the case, and because wine distribution is so erratic, having a supply of a wine you like ensures at least a modicum of continuity. But unless the wine is to be stored for ten years or so, or unless you live in Los Angeles or some other equatorial village, you simply do not need elaborate storage facilities.

Some 95 percent of the world's wine is consumed within the year after it's made. Of the rest, probably 95 percent is consumed within five years after it's made. Wine is tougher than we think; it survives often horrendous journeys from the winery across the seas to the wineshop, frequently punctuated with long siestas in some whiskey distributor's superheated warehouse.

A constant temperature of 55 degrees is much to be admired—in a chateau cellar or in the basement of a great restaurant. In a home where the wine will be drunk over the next few months, or even years, the temperature can be 65 or 70 degrees or whatever—so long as it doesn't change drastically or often.

Vibration can be a problem, but few of us live next to busy truck routes. Some years ago, a chap was trying to build a serious wine collection in a cellar directly over the BMT subway line in Brooklyn Heights. The whole place shuddered with the passage of each train. The only advice anyone could give him was to drink young wines or to move.

Or—and here is at least a partial solution for serious city collectors— you can turn your wines over to a storage company. For a price (so much a month for each case), they will hold wines in temperature-controlled warehouses. The average fee is about $1.30 a month per case, depending on the number of cases, which is far less than the cost of an elegant wine cellar. The difference is that you can show your friends your cellar, or your air-conditioned hall closet, but you can't easily take them to your warehouse.

JUNE, 1987

## STILL LOOKING FOR NEW VINEYARDS TO CONQUER

One of the most remarkable figures in the French wine business, Jean-Michel Cazes, retired in 2000 at sixty-five. Sort of.

In 1986, when the conventional wisdom held that most French businessmen were still trapped in the dark ages, Mr. Cazes was already well ahead of the crowd. With his friend Claude Bebear, Mr. Cazes was running a successful insurance agency in Pauillac, in the center of the Bordeaux wine country. He was also the owner of Château Lynch-Bages and other wine properties. Mr. Bebear was the president and managing director of

AXA, a huge insurance group. The two had met in the 1950s when they were students and rugby players in Paris.

AXA had decided to get into the wine business, and Mr. Bebear asked his friend about buying Cantenac-Brown, a well-known wine chateau in Bordeaux. Mr. Cazes told him to do it. But another group, Compagnie du Midi, outbid AXA. Mr. Bebear waited and then bought Compagnie du Midi.

In the meantime, he asked Mr. Cazes to head a new subsidiary called AXA-Millésimes, formed expressly for buying and developing wine properties. Under French law, insurance companies must invest part of their assets in French real estate. In the 1980s, as the Bordeaux wine trade came out of a decade's slump, the big insurance companies began to move. As a French newspaper said at the time: "They bought Paris, then they bought the forests; now they want to buy Bordeaux."

Working together, Mr. Bebear and Mr. Cazes assembled a top-line portfolio, which includes Pichon-Longueville Baron, Cantenac-Brown, Suduiraut, and Petit-Village chateaus in Bordeaux and, moving away from France, the Quinta do Noval, a Port producer in the Duoro Valley in Portugal, and the Disznoko estate in Hungary, which makes Tokay.

They are always attracted by high-quality properties.

Mr. Cazes took on AXA-Millésimes while continuing to manage his family's holdings, which include Lynch-Bages, Château Les Ormes de Pez in St.-Estèphe, and Villa Bel-Air, a chateau in the Graves region, south of Bordeaux. Of course, there is still the insurance agency in Pauillac and, almost next door to Lynch-Bages, where he lives, Château Cordeillan-Bages, once a minor wine estate and now an elegant small hotel and restaurant, with its own vineyard and a wine school.

Lynch-Bages is the jewel in the Cazes diadem. In the famous 1855 classification of the sixty top Bordeaux chateaus, it was placed in the fifth, or lowest, group. Under Mr. Cazes, Lynch-Bages has become one of the finest of the rated chateaus. Today it is informally considered among the best of the original sixty.

Mr. Cazes's retirement implied no change in his frantic pace. It means only that he turned sixty-five, and AXA said he had to give up that side of his life. His friend Claude Bebear had to go, too.

The Cazes family was not born to the chateau life. They came from the Pyrenees, near the Spanish border, farmers and shepherds for centuries. The first Cazes to arrive in the Bordeaux region, in the mid-1800s, was named Jean. His second son, Jean-Charles Cazes, became a baker, worked for a bank, and then opened his insurance agency in 1927. In 1933 he became a tenant farmer at Lynch-Bages, buying it outright five years later. In 1940 he bought Les Ormes de Pez.

André Cazes, his son and Jean-Michel's father, took over the insurance business in 1945, after five years as a prisoner of war of the Germans. In 1949 André Cazes was elected mayor of Pauillac, a post he held for forty-two years.

Jean-Michel Cazes planned a life away from Bordeaux and wine. After graduating from France's M.I.T., the École Nationale des Mines in Paris, he spent a year studying petroleum engineering at the University of Texas in Austin. Back in France and in the air force, he was exposed to computers for the first time, which led to twelve years with I.B.M. in Paris. By 1973 his father needed help in Pauillac, and the younger Mr. Cazes went home. They began a fifteen-year renovation program at Lynch-Bages and Les Ormes de Pez.

They decided, too, to make Lynch-Bages a top-quality wine. A second label was created, Haut-Bages-Averous, and henceforth only the best vats of wines were used in Lynch-Bages. The rest went, as it does today, into Haut-Bages-Averous.

For more than a century, Bordeaux was dominated by a small, clannish, and powerful group of merchants who bought the wine from each chateau, aged it, and bottled it in their own cellars on a section of the Bordeaux waterfront known as the Quai des Chartrons. The Chartronnais, as they were called, set prices, lent money, and called in mortgages at will. One of the first to break the stranglehold of the Chartronnais was Baron Philippe de Rothschild, at Château Mouton-Rothschild, who insisted on bottling and selling his own wine. Alexis Lichine is often credited with dragging the old conservatives into the 20th century with his use of marketing and promotion.

Mr. Cazes, partly with AXA's help but mostly on his own, has emphasized the global dimensions of the Bordeaux wine trade. Thanks in part to his example, the younger Bordelais are as comfortable flying off to Bangkok for a sales meeting as their fathers were dashing off to London for a couple of Savile Row suits.

Mr. Cazes still has a seat on the AXA-Millésimes board. He also owns a wine wholesaling business and part of a company that makes wine management software. He is grand master of the Commanderie du Bon-temps, a chateau owners promotional group. And he is looking at vineyard and winery investments around the world, including other properties in Bordeaux.

But he insisted: "I am going to cut back. Really. I'm going to retire from my insurance business next year, and my wife and I are going to travel more."

"Anyway," he added, almost guiltily, "the Internet, e-mail, and a good connection allow everyone to keep in touch from anywhere." Some retirement.

JUNE, 2000

## KEEPING BENMARL

Benmarl, a vineyard, winery, and modest social experiment in the Hudson Valley, celebrated twin 20th anniversaries in 1991: its own as a commercial venture, and that of its Société des Vignerons, one of the least-known, but hardiest, wine clubs in the country. It also looks ahead to an uncertain future.

This pioneer winery covers seventy-two steeply sloped acres on the west bank of the Hudson in Marlboro, New York, ninety minutes north of the George Washington Bridge. It was the model and inspiration for the dozen or so Hudson Valley wineries that followed. But real-estate developers have been eyeing the Benmarl property and unless its owner, Mark Miller, who is now seventy-two years old, can find a way to preserve it as a vineyard, Benmarl could disappear.

The vineyard on the site is the oldest continuously operated vineyard in the United States, dating to colonial times, Mr. Miller said. He bought it in 1957 when it was little more than a couple of battered farm buildings and thirty-six acres of tired vines whose Dutchess grapes were used to make grape juice and jelly.

The son and grandson of prosperous Oklahoma cotton growers, Mr. Miller took a different path than his forebears. After a stint as a Hollywood costume designer, he became a sought-after illustrator whose work appeared regularly in the *Saturday Evening Post* and *Collier's* magazines. He grew a few grapes in his backyard in suburban Connecticut and made some wine. Benmarl—*ben* is Gaelic for "hill," and *marl* describes its mixture of soils—started as a weekend project. "A place to plant a few vines, with a comfortable little summer house" is the way he described it later.

In the early years, that's just what it was: a weekend diversion for Mr. Miller and his family. Then came seven years, 1960 to 1967, when the Millers lived in Burgundy, in France. Mr. Miller drew and painted, and also fell in love with the wine business.

Back in this country, he resolved to devote his life to winemaking in the Hudson Valley. Early on, he decided that he would concentrate on grapes from hybrid vines that traced their parentage to both elegant European grapes and hardy American varieties capable of withstanding the rugged winters of the Northeast. The principal vines in Benmarl's vineyards are seyval blanc, vignoles, and verdelet for the white wines, and chelois, baco noir, chancellor, and foch for the reds.

In 1971, to promote his vineyard and its wines, Mr. Miller founded his Société des Vignerons. Each vigneron (the word is French for "wine-grower") owns rights to two producing vines in the Benmarl vineyard. For this privilege he or she pays an initiation fee of $250 and an annual maintenance charge—dues, actually—of $150. In fact, while the vine rights can be sold, donated, and deeded to heirs, they are mostly symbolic and represent no land ownership.

What they do represent is the right to buy Benmarl's widely respected wines at a discount, as well as membership in a group of wine enthusiasts from all walks of life who take a proprietary interest in Benmarl and its future. They come each year to pick grapes, picnic among the vines overlooking the Hudson, and participate in a lively round of social events that include concerts, clambakes, wine classes, dinners, and parties to mark the seasons of the wine calendar.

These activities, as well as the daily routine at the winery and in the vineyards, are chronicled in a monthly newsletter edited by Mr. Miller and distributed to all the members of the society. Mostly the newsletter is cheery, in the avuncular style of an alumni letter: Don't forget to pay your dues, to contribute to the class funds, to come to the reunion. But lately the Benmarl letter has sounded a more ominous note.

At its peak in the early 1980s, the Société des Vignerons had almost fourteen hundred members. The figure is down to around eight hundred, and this poses a serious problem for Benmarl because traditionally the vignerons have bought up to 90 percent of the vineyard's production of ten thousand cases a year. Beyond the immediate financial problem is the future of Benmarl itself.

Mr. Miller's only logical successor, his son and former Benmarl wine-maker, Eric Miller, is totally involved in his own winery, the Chaddsford Winery in Chadds Ford, Pennsylvania. There are potential buyers, but they want the property for development.

"Realizing that at the age of seventy-two I am the last member in resi-dence of the two families (the Caywoods and the Millers) which have been the stewards of this remarkable vineyard for over two hundred years," Mr. Miller wrote to the society's members recently, "makes me aware that it would be wise to take precautions to insure that Benmarl can continue to make wine and also serve as a cultural and historic landmark into the in-definite future." His plan: to try to donate the property to a university, per-haps Cornell. But, he notes, such a gift has to be accompanied by a substantial endowment of operating capital.

Mr. Miller proposes that members of the society buy the 80 percent of the company that he owns and donate the shares to a Benmarl Heritage Trust he proposes to establish. The trust would dedicate the Benmarl shares to a beneficiary, perhaps Cornell, while revenue from the sale of shares could be used for current operations. The donors would receive immediate income-tax deductions and participate in any Benmarl earnings for life or for an agreed-upon period.

The revenue raised from the sale of shares, Mr. Miller said, would also be used to inaugurate educational programs, to increase membership in the Société des Vignerons, and to develop new markets.

The trust is a relatively long-range project. Benmarl's more immediate goal is to build up membership in the society. "A substantial part of our group has matured and is no longer active enough to generate the much-neded funds to run the vineyard and produce our winning wines," Mr. Miller wrote. "Unless we all take a stand, what we have today could be a dream of yesterday."

APRIL, 1991

## A TOAST TO BEAUNE

B eaune is a wine town. It is inhabited by wine shippers and winegrowers, and it is invaded regularly by wine buyers and wine tourists. It
stands on millions of gallons of wine in its cellars and is surrounded by
thousands of acres of wine grapes. It would be possible to visit Beaune and
ignore the wine, but it would not be easy. It would be like visiting Las Vegas and ignoring the casinos.

Beaune is the wine capital of Burgundy, and the embodiment of the
Burgundian spirit. This is important: Burgundy is actually more a state of
mind than a place. Even the French, who hunger after order as only a
people who have no hope of achieving it can, have not tried to define Burgundy's borders. Burgundy as a political entity ceased to exist after the death
of Charles the Bold, the last of the powerful Valois dukes of Burgundy, in
1477.

Today's Burgundy is a ruddy face under an old béret, the sound of a
shotgun on a chill fall morning, the dark, foresty taste of the bird it killed.
It is the 12th-century abbey at Pontigny, where Thomas Becket sought
refuge; it is the tranquil Canal de Bourgogne, defined by rows of poplars
standing with military precision along its banks. It is good food in fine
restaurants, and it is the wines whose sonorous names are almost as delightful as the wines themselves: Montrachet, Monthélie, St.-Véran, Savigny,
Santenay, Richebourg, Romanée, Pommard, Puligny. It is the tiny villages
that produce the wines, most of them consisting of only a few houses, in
spite of their international renown—such places as Aloxe-Corton, Meursault, and Nuits-St.-Georges, with their little museums, cellars for tasting,
modest restaurants, and cafés where feet can be rested and thirsts slaked
with beer, not wine. Wine only makes you more thirsty.

At the center of all this, about three and a half hours from Paris by the
Paris–Marseilles *autoroute,* is Beaune, with its narrow cobbled streets and old
houses, its ancient trees and hidden gardens, its steep roofs and handsome
ramparts. Beaune also with its fleets of tour buses and their sheeplike pas-
sengers, its overpriced gift shops and its endless signs: "Stop Here and
Taste." "Direct Sales of Wine." "Visit Our Cellars."

The pace slows in autumn. The vines and the leaves on the trees turn
red and yellow, and if the year promises a good harvest, there is a sense of

excitement in the air. Many, but not all, of the summer's tour buses are gone, and there are rooms to be had in the hotels and at tables in the best restaurants.

"One must have time to live," the French say, and there is time in Beaune in autumn. Time to visit the cool dark cellars, to taste the wines, to stroll around the ramparts and through the little wine museum installed in one of the former residences of the dukes of Burgundy. There is time to wander through the church of Notre Dame, with its deep porch, typical of Burgundian architecture, and its superb late-15th-century tapestries depicting the life of the Virgin. And there is time, above all, to savor the riches of the Hospice that was founded by Nicholas Rolin, Chancellor of Burgundy, in 1443.

The Hospice, which has been a working hospital since its founding, is actually two major buildings, the Hôtel-Dieu and the Hospice de la Charité. The Hôtel-Dieu is Beaune's grandest attraction, with its perfectly preserved Burgundian-Flemish architecture and priceless art collection. Its somber stone façade is surmounted by a vast, steeply sloping roof, tiled in a lozenge pattern of red, yellow, black, and white. Inside, its most striking feature is the Grand'Salle, or Paupers' Room, 160 feet long, still displaying the original 15th-century furnishings, among them twenty-eight red-canopied and red-curtained beds used by the indigent patients of five centuries ago. The Hôtel-Dieu is open all year long, and there are guided tours of the Grand'Salle, the old pharmacy and kitchen, and the museum full of art treasures.

The masterpiece of the museum—and, indeed, one of the masterpieces of the whole canon of Western art—is the polyptych of the Last Judgment by the Flemish artist Roger van der Weyden that Chancellor Rolin commissioned to hang in the Grand'Salle. Also on display are superb tapestries, among them the covers once used on the beds in the Grand'Salle. Elsewhere in the Hôtel-Dieu can be seen some of the habits worn by the Dames Hospitalières, the nuns who ran the hospital until 1961, several collections of rare pewter, and the visitors' book of the hospital, opened to the signature of Louis XIV.

But, like most places in Beaune, the Hospice is also concerned with wine. It owns pieces of vineyards all over Burgundy, and each year, on the third weekend of November, it sells its wine at auction. For prominent wine buyers throughout the world, attendance at the auction, at least every two years or so, is *de rigueur.*

The wines, which are only two months old at the time of the auction, are available for tasting on Saturday morning in the Hospice cellars and are auctioned the next afternoon in a drafty, chilly hall in the center of town. The prices are unconscionable because the high bidders like to get the publicity, and besides, the money goes to charity.

The Hospice auction weekend is the most exciting in Beaune's calendar but not necessarily the best time for a visit, unless you are in the wine business. All the hotel rooms and all the best restaurants are booked long in advance; waiters, clerks, and guides are harried; and it is difficult not to feel left out when everyone else in town is dashing off to dinners, parties, and tastings.

But a happy few will not be left out. They will have received invitations to one of the grand dinners given at the Clos de Vougeot. There are seventeen a year, eight in the fall and eight in spring, as well as one on the feast of St. Vincent, patron of winemakers, in January. The most important are the three that coincide with the auction at the Hospice. They are called Les Trois Glorieuses—The Three Glorious Evenings.

The Trois Glorieuses and other festive dinners are the preserve of the Confrérie des Chevaliers du Tastevin, a promotional group created during the Depression to promote Burgundian wines. Its members rig themselves out in mock-medieval robes, give big dinners, drink good wines, and manage to have a lot of harmless, if fattening, fun.

The Chevaliers own the Clos de Vougeot, a former monastery a few miles north of town in the middle of the vines. The main hall of the Clos seats 550 people, and that is how many are invited to each Confrérie dinner. The dinners are for members of the Confrérie and occasionally their friends; even more occasionally, the visitor can wangle an invitation through the management of his hotel.

Imagine the scene: It is a cold November night. A half dozen Americans are sitting around the lobby of the Hôtel de la Poste, dressed in evening clothes, sipping Champagne. They are trying to look as if they always sit around the lobby of the Poste in evening clothes drinking Champagne. Actually, they are nervous.

The Americans think that a dinner at the Clos de Vougeot is serious business. They adjust their ties too often and discuss the vintage self-consciously. At the appropriate time, they climb into their rented cars and head north. They get a huge surprise. There is singing, shouting, much backslapping and laughter. Wines are rarely mentioned and no one seems to care whether or not any of the Americans knows the difference between a Beaune Clos des Mouches 1964 and a jug of Spanish plonk bottled last week. The most profound wine comment is a sip, a frown, a big smile, a dig in the next man's ribs, and a "Hey, hey, not bad. Not bad at all, eh, *mon vieux?*"

The truth is that, like much in and around Beaune, the dinners at the Clos de Vougeot have only one purpose—to sell wine. But the sight of the Clos, dark against a full moon, with a blaze of light spilling into the court-yard as the guests enter—this is very stylish selling.

But as for buying wine in Burgundy, it is probably better not to, unless you know what you are doing. Knowledgeable buyers who have a supplier they know and trust can do well, but the casual tourist must rely on the shops around the Place Carnot in Beaune, where prices are often higher than they are in New York. Better to taste around, try different wines with meals, then search them out back home.

It is likewise better not to visit Beaune without reservations. Not just hotels, but also restaurants are apt to be booked months in advance for the best fall weekends. The restaurants are not to be missed—Burgundian food can be as glorious as Burgundian wine. It is, traditionally, hearty fare, the food of farmers hungry after long days in the fields and vineyards. Snails cooked in their shells, *coq au vin,* an enormous variety of sausages and salmis, terrines of pork and veal, all kinds of game in season—these are the bases of the best Burgundian cooking.

It is not a bad idea to write to Beaune shippers for appointments to see their cellars. Some, though admittedly few, offer no tours or tastings; others are particularly hospitable to visitors who write in advance. In any event, never hesitate to take wine tours or to visit vineyards because of a fancied or real lack of knowledge of the subject. The Beaunois have been spreading the word about their wines since Pliny wrote about the Gauls drinking wine at Auxerre in the 5th century. The present-day Gauls will be happy to tell you whatever you need to know, and to sell you a couple of bottles as well.

OCTOBER, 1981

## THE BRASH PRINCE OF
## WINE COUNTRY

A rooster tail of dust billowing out behind it, the gold Lexus climbed into the rugged Mayacamas Mountains. "Don't be nervous," said Jess Jackson, pulling deftly through another hairpin turn. "I know this country."

He should; it's the Gauer Ranch, some seventy-five miles north of San Francisco, one of the last great landholdings in Sonoma County: more than five thousand acres of mountains, meadows, and, yes, vineyards. And Jess Jackson owns it all.

"I often read about the Gallos," he said as he drove along, ignoring a three-thousand-foot drop just inches from the car. "Everyone talks about how much land they own up here. Well, we own a lot more than they do."

What Mr. Jackson owns is nothing less than the fastest-growing and arguably most successful winemaking business to appear in California in twenty-five years. From a few acres of grapes in the early 1970s, he now controls more than a dozen wineries—including his flagship, Kendall-Jackson—stretching from all the choice growing regions of California to Europe and South America. Though still dwarfed by the E. & J. Gallo Company, the industry giant, Mr. Jackson's empire is now among the top ten wine businesses in the country, with sales of about $200 million in 1996 from Kendall-Jackson alone, according to Jon Fredrikson, an industry economist. The smaller wineries produced another $30 million, Mr. Fredrikson said.

In the process, he has managed to change the way fine wine is made and marketed in the United States, leaving many in the business to marvel at his highly lean, efficient, and integrated operation, which includes a barrel-making business and a distribution company. At the same time, the pugnacious Mr. Jackson, an experienced lawyer, has become the object of envy as well as admiration, and has generated a bumper crop of speculation and controversy.

Mr. Jackson's story combines the good fortune of a novice who has stumbled onto something new with the calculating eye of a master marketer who has found his magic niche.

An early mistake in the fermentation process of his Vintner's Reserve chardonnay created a surprise hit whose slight sweetness has inspired others to offer similar products. The competition also followed in his footsteps when Mr. Jackson carved out a new middle market for fine wines after carefully analyzing the existing price structure.

He has been largely alone, however, in bringing a lawyer's hardball style to the traditionally genteel wine business. Mr. Jackson took advantage of an industry downturn in the late 1980s to snap up unprofitable vineyards and hire away well-regarded winemakers. And he has been involved in several high-profile legal battles, even taking on Gallo.

Now, as his business is poised for still more growth, the seventy-year-old Mr. Jackson is clearly a major force to be reckoned with, even though, as he proudly says, he is still not part of "the old boy system."

"Jess came into the business as a real estate lawyer, and he still thinks like a lawyer," said one winery owner who has known Mr. Jackson for over twenty years. "That's why wine people can't figure him out; they're not lawyers."

Rodney Strong, the founder of another Sonoma winery, Rodney Strong Vineyards, added, "He's not an insider in the wine business, but he's taught the insiders some valuable lessons about making and selling wine."

The original Jackson vineyard was a weekend hobby. By the late 1970s, Mr. Jackson, then a San Francisco lawyer, was selling grapes to several wineries. "I never planned to get into the wine business," he said recently, "but there was a grape glut around 1978 and I couldn't sell what I had grown. So I had it made into wine."

His first vintages were called Château du Lac. By 1982 the name had been changed to Kendall-Jackson—Kendall is the surname of Mr. Jackson's first wife—and a winery was under construction.

While making his 1982 chardonnay, Mr. Jackson ran into a problem. At a crucial point, the fermentation stopped before all the sugar in the grapes was converted into alcohol and carbon dioxide. Efforts to restart the process failed, and the wine was bottled with the unfermented sugar—just a touch—still in it. Vintner's Reserve was an instant success, and it has stayed that way. The reason, experts say, is that the new product legitimized the sweet taste of jug wine, attracting an army of drinkers willing to pay a few dollars more for the status of a better label.

From 18,000 cases that year, total Kendall-Jackson shipments rose to an estimated 2.6 million cases in 1996, with Vintner's Reserve accounting for

64 percent. No threat to Gallo, which produced an estimated 55 million cases in 1996, but impressive for a fourteen-year-old upstart.

"Jess is a genius at calculating price points," one of his competitors said. "In his early days, wineries like Sebastiani were going for the low end—$3, $4 a bottle. Others, like Château St. Jean, were developing the $9 category. Jess saw a space between the two and filled it."

Mr. Jackson concurred. "We sold out the '82 vintage in six months and price was certainly part of it. We started just under $5, then we raised it twelve times in twelve years." Vintner's Reserve now retails for around $12.

Besides Kendall-Jackson, now based in Santa Rosa in Sonoma County, Mr. Jackson's holdings include eleven wineries in California, a winery and vineyards in Italy, a leased winery in Chile, and two square miles of vines in Argentina. The wineries other than Kendall-Jackson are grouped under the name Artisans and Estates and produce small batches of higher-priced wines.

Still to come are five new wineries in California.

"We are always looking for good properties," Mr. Jackson said.

Tall and rangy, with a shock of silver-white hair tumbling into his eyes, Jess Jackson bears a resemblance to Charlton Heston. Invariably dressed in jeans and boots, his face tanned from long exposure to the California sun, he looks every inch a man of the soil.

In fact, he is a city person, born in Los Angeles and raised in San Francisco. As a boy, wine was a part of his life. "San Francisco had a rich wine culture," he said. "I had an Italian uncle who made wine, and we drank it regularly. During the war, when they needed people in the vineyards, I picked grapes."

Mr. Jackson went to law school at the University of California at Berkeley, paying his way by working on the San Francisco docks and as a Berkeley policeman. While still a student, he did research for the late Pat Brown, then California's attorney general and later governor. As a young lawyer, Mr. Jackson worked for the state, specializing in real-estate law, invaluable training for a man who would become the state's busiest vineyard buyer.

Indeed, in California's wine business, which nurtures its laid-back image, Jess Jackson's drive and combativeness stand out. He garnered considerable attention years ago, for example, in a legal battle with Jedidiah T. Steele, his longtime winemaker. At the heart of the dispute was the accusation by Mr. Jackson that Mr. Steele had revealed "trade secrets" about the process for making Vintner's Reserve shortly after leaving to start his own business. When Mr. Jackson stopped making severance payments to him, Mr. Steele sued, and Mr. Jackson countersued.

The fracas further separated Mr. Jackson from the wine establishment. Robert Mondavi and other prominent winemakers testified on Mr. Steele's behalf, saying that there was nothing really secret about the process. A ruling against Mr. Steele, they said, could stifle the cooperation that they maintained had enabled the California wine industry to make great strides in a relatively short time. The court ordered Mr. Jackson to pay Mr. Steele the balance of the severance agreement, but also found that Mr. Jackson's winery was entitled to keep the process to itself.

As his business expanded, Mr. Jackson roiled the industry's seemingly placid surface by hiring away some of the competition's best people, then roiled it again when he announced that the federal government's labeling regulations, accepted by the rest of the industry, needed to be tightened— a grandstanding move, critics said. (The Government is still reviewing his petition.)

Currently Mr. Jackson is locked in a legal battle with Gallo over trademark infringement, accusing Gallo of copying a Kendall-Jackson label for its Turning Leaf line. This is his second courtroom tussle with the Gallos. In 1986, in his previous life as a lawyer, Mr. Jackson represented Joseph Gallo Jr. in his unsuccessful attempt to wrest a share of the company from his older brothers, Ernest and the late Julio.

Kendall-Jackson's dramatic expansion at a time when other wineries were experiencing serious economic problems gave rise to some equally dramatic talk about the company's financial circumstances. Mr. Jackson brushed off his critics with an impatient wave of the hand. "We are in excellent financial condition," he said, "and it gets better all the time. Why, we are the Bank of America's favorite winery."

Mr. Jackson's hiring of a Los Angeles financial public relations firm a few years ago gave rise to talk that he was about to take the company public. "We reserve that option," Mr. Jackson said, "but we have absolutely no plans to do it now."

Jean-Michel Vallette, who watches the wine industry as a partner in Hambrecht & Quist, a San Francisco investment bank, agrees with the upbeat assessment. "In the early days, Jess clearly needed outside help and was certainly highly leveraged," Mr. Vallette said. "But today he can attract the highest-quality lenders and get the most attractive financing terms. People have always underestimated just how profitable Kendall-Jackson is."

Mr. Jackson releases no precise figures but doesn't argue with an industry estimate that Kendall-Jackson has profits of about $40 million a year just from Vintner's Reserve.

At last count, at least fourteen members of Mr. Jackson's family were working in the business. His second wife, Barbara Banke, also a lawyer, is president of one of the wineries.

The Jacksons live in a small frame house that stands in stark contrast to the palaces owned by some of his competitors. The house is just across the road from the winery, where he keeps his office. "I prefer to put my money in the business," Mr. Jackson said, sounding something like a man on a mission. "I want to broaden the consumption base" of wine, he said, "and that means educating people. Wine is made out to be foreign, as opposed to beer, which is made to be American. We've got to make it more American. It's not going to blossom in American culture until we take the mystique out of it."

JANUARY, 1997

## WHO NEEDS VINTAGE CHARTS?

About a decade ago, Bruno Prats, then the owner of Château Cos d'Estournel, in St.-Estèphe, declared that there would be no more bad vintages of wine in Bordeaux. The height of arrogance, I thought at the time, it being a characteristic not unknown among the Bordelais. I was wrong.

A bit of an overstatement, perhaps, but, essentially, Mr. Prats was right on the mark. Great wine may still be elusive, but rarely now does a year go by that doesn't produce good wine, even in marginal regions like Bordeaux, where the weather is as risky as a dot-com stock.

The fact of the matter is that in the cellar and the vineyard, the wine-makers of the world have rendered the vintage chart obsolete. For the uninitiated, the vintage chart tracks various categories of wine over a period of a year. Most vintage charts use numerical ratings; some add a code

indicating whether the wines are too young to drink, ready to drink, or seriously past their prime.

Over the years, every year, I have produced vintage chart after vintage
chart, always adding enough qualifications and caveats to make the reader
wonder why I bothered in the first place. I am not alone. Each year, Robert
M. Parker Jr. publishes a vintage chart of daunting thoroughness. This year's
has twenty-eight separate categories. Even so, he warns: "This vintage chart
should be regarded as a very general overall rating slanted in favor of what
the finest producers were capable of producing in a particular viticultural
region. Such charts are filled with exceptions to the rule: astonishingly
good wines from skillful or lucky vintners in years rated mediocre, and thin,
diluted, characterless wines from incompetent or greedy producers in great
years."

Vintage charts assume a high degree of homogeneity: Northern California, the Central Coast, the South Central Coast. But can a single rating
for Northern California adequately cover the reaches of northern Mendocino, the furnacelike valley floor of Calistoga and the fog-shrouded hills of
Carneros? And what about the Amador County vineyards, far to the east?
In Sonoma County, the wines of Dry Creek are different from those of the
Alexander Valley, which are not the same as those from the Russian River
or the Sonoma Coast.

The first vintage charts came from France and were compact cards with
listings for Bordeaux, Burgundy, the Rhone, Champagne and the Loire.
Today, superior wines are made in Alsace, the Languedoc, in Roussillon,
and the Southwest, often when wines from more traditional regions of
France are less than exciting.

Can any single vintage chart do justice to Italy, where there are a hundred different wine regions, where remarkable wines might be made in the
foothills of the Alps and in Sicily in the same year that mediocre or poor
wines come from Tuscany and Umbria?

A chart might report conscientiously that 1997 was a good year in
Spain. Where in Spain: Penedes? Rioja? Navarra? Priorat? Ribera del
Duero? And what of the newest wine sources—Chile, Argentina, Australia,
New Zealand, and South Africa? Don't they deserve space on the charts?
And even if they got it, what chart could show their growing diversity?

Both Chile and Argentina are developing new wine regions far to the
south of their traditional vineyards; Australia's Margaret River region is

more than two thousand miles from its Hunter Valley, and South Africa makes different wines in Hermanus from what it produces in Stellenbosch.

A vintage chart might advise that the 1998 Bordeaux will need several years of bottle aging after they are bottled and shipped. But a consumer finds the shops already filled with 1998s that, he is told, are ready for immediate drinking. The chart, of course, is concerned only with expensive wines, which spend several years aging before they are sold. It ignores the hundreds of lesser wines that often are uncomfortably close to their better-known siblings in quality.

The vintage chart speaks to wine regions at a time when winemakers—and consumers—are increasingly concerned with *terroir,* the uniqueness of small plots of land. Some vintners now produce five or six separate wines from half a dozen small, contiguous vineyards, while others make two or three different wines from the same vineyard.

There are highs, like 1945, 1961, and 1982 in Bordeaux and 1994 in the Napa Valley, and lows, like 1991 or 1992 in Bordeaux and Tuscany, but not one year when some decent Bordeaux wine was not produced. But it's unlikely that we will ever see vintages like 1963 and 1965, dreadful years in Bordeaux that resulted in dreadful wine. There will always be years in which nature doesn't cooperate, but even in those years, some decent, drinkable wines will be made.

Everywhere, hardier rootstocks, better grapes, limited yields, and severe grape selection at harvest time have increased quality. So, too, have new organic methods of pest and disease control, and new planting techniques. Now, NASA satellites tell growers where to plant and which efforts have been the most successful. This is not quite a golden age of wine; in spite of everything, there will always be dull or poorly made wines. But those are not the problem of the vintage. And competition, which is keener than ever, makes it difficult for plonk to survive in the market.

Vintage charts are not entirely passé. Wine can be confusing, and the chart does offer some general information, as Mr. Parker, the critic, suggests. And, of course, in the auction rooms where collectors haggle over the 1982 Pétrus and the 1994 Musigny, vintage charts can be a handy reference tool. Elsewhere, though, vintages are increasingly irrelevant.

In the past, letters would begin to arrive about a year after the last vintage chart appeared. When, the writers inquired, would the next chart appear? No longer. A lot of people must be deciding for themselves.

In the final frames of the film *Little Caesar,* Edward G. Robinson snarls: "Is this the end of Rico?" To my way of thinking, that sacred talisman of the wine buff, the vintage chart, is just as dead as Rico was when the screen went to black.

FEBRUARY, 2000

## THE 1855 RATINGS

The 1855 classification of the wines of Bordeaux is probably the most important wine list ever written. Yet few people, even among smart wine drinkers, know why it's important or even why it was put together in the first place.

In 1851, Britain mounted a spectacular exhibition in the Crystal Palace in London to show off its industrial might. Not to be outdone, the French decided that they too would show the world what they could do. Thus, in March 1853, barely three months into France's Second Empire, a great Universal Exposition was decreed. To be held in Paris two years hence, it would celebrate a return to French grandeur, the wonders of modern (French) industry, and, not the least, the glorious, if faintly illegitimate, apotheosis of Louis Napoleon, citizen, into Napoleon III, Emperor of France.

It was a time not unlike our own. Money was plentiful, peace reigned, and thousands of workmen, driven by Baron Haussmann, the Robert Moses of his day, were carving Paris into an urban jewel.

With chefs as popular as courtesans and the Rothschilds eying wine chateaux in Bordeaux, it was decided early on that food and drink would play important roles in the 1855 Exposition. The Bordeaux Chamber of Commerce was asked to choose the wines and see that they got to Paris in time and were displayed properly. The chamber invested licensed wine brokers in Bordeaux with the task of selecting and classifying the wines to be shown in Paris. They selected sixty-one out of the hundreds of possibilities available to them and divided them into five categories called *growths.* The brokers, known as courtiers, were (and still are) go-betweens, working for both the wine producers and the wine merchants or shippers. Truly expert tasters, the brokers advise the merchants what to buy and how much to pay.

For the chateau owners, the brokers find merchants who will buy their wines. Many nineteenth-century brokers grew wealthy on their commissions, becoming chateau owners and shippers themselves.

In seeking wine for the Paris show, the brokers did some blind tasting, but for the most part, classifying the wines had little to do with tasting skills and almost everything to do with past prices. Some wines—many, in fact—had over the years consistently fetched higher prices than the others. Thus, chateau owners, by plotting the sales of their wines and others through five or ten vintages, could easily see which wines were consistently the best or, at least, had been judged the best by the market. The sixty-one chateaux that had done the best over the years made it to the honors list.

The emphasis on price helps to explain why no wines from the St.-Émilion-Pomerol region ever made the list. Good as they were, they never commanded the high prices wines from the Médoc and Graves did.

Classifications were nothing new, even in the years leading up to 1855. One of the earliest had been made more than two centuries earlier, in October 1647. It rated not wines but wine-producing communities within the Bordeaux region. The results showed that the best wines—the wines that fetched the highest prices—were not that different from what they were in 1855 or, for that matter, what they still were in 2001. The highest rating went to the Graves and the Médoc for the reds, and to Sauternes, Barsac, Preignac, and Langon for the whites. A survey done in 1745 showed Châteaux Margaux, Lafite, and Haut-Brion selling for up to 1,800 francs a ton, while chateaux like Gruaud-Larose and Beychevelle brought only 400 to 600 francs.

Four decades later, in 1787, Thomas Jefferson, the ambassador to France, made his own classification, selecting Latour, Haut-Brion, Lafite, and Margaux as his favorite Bordeaux wines. Today, as they were in Jefferson's time, these four plus Mouton-Rothschild, are still the top five, the so-called first growths.

In the nineteenth century, before the classification was generally accepted, owners often lobbied the local agricultural authorities for a boost into a higher category. But the only elevation since the creation of the 1855 list took place in 1973, due to Baron Philippe de Rothschild's long battle to have Château Mouton-Rothschild moved up from a second growth to a first. After twenty years of pleading and politicking, he pulled it off. He celebrated by putting a Picasso from his private collection on the new label.

Why has the 1855 classification never been modified? No one really knows. It certainly needs it. Chateau owners die, managers move on, vineyards are neglected. Other chateaux improve. Lynch–Bages, in Pauillac, was a fifth growth in 1855. Today it consistently rates either with the second growths or the first in tasting competitions. In spite of the remarkable consistency of so many Bordeaux wineries over the years, ratings need to be updated regularly.

The history of the 1855 ratings is the subject of a recent book, *1855: A History of the Bordeaux Classification* by Dewey Markham. Working in the musty libraries of Bordeaux for almost four years, Mr. Markham unearthed a storehouse of wine history and folklore, including the saga of Monplaisir Goudal. M. Goudal was the manager of Château Lafite at the time of the 1855 classification, and he worked unendingly, records show, to enhance the chateau's reputation—and his own. To give the 1846 Lafite, one of the wines exhibited at the fair, a bit of age, M. Goudal had fifty bottles of the wine sent round the world to, as Mr. Markham said, "subject the wine to the accelerated aging that such sea voyages provoked."

Accelerated aging? Clearly, the man had his own classification.

APRIL, 1998

# CHAPTER 5

# PERSONALITIES

No field of endeavor, not medicine, not diplomacy, not the fine arts, not even literature attracts to it as remarkable a group of individuals as does wine. Start with Philippe de Rothschild, a poet, a scholar, a film producer, an automobile racer, a war hero, and a major art collector as well as a winemaker. His daughter, the Baroness Philippine, who carries on his work and his style at Château Mouton-Rothschild, is no less dynamic than he was.

Robert W. Parker Jr. is the most influential wine critic in the world. André Tchelistcheff was America's most important winemaker for almost sixty years. Dick Grace doesn't drink but makes Grace Family Vineyards cabernet, one of the most sought-after of all American wines. Alexis Lichine, the Russian-American who loved France and hated the French, taught his adopted country how to drink wine, and Sandy McNally, a Lichine protégé, became the first modern wine auctioneer.

Odette Pol-Roger was a "grande dame" of Champagne who supplied her friend Winston Churchill with Pol-Roger every month for almost twenty years until his death, and Len Evans, who is Australia's ambassador to the wine world, really wanted to be a professional golfer.

Singly or together, they are serious wine professionals and, to me at least, utterly fascinating people.

## PHILIPPE DE ROTHSCHILD: A MAN WORTHY OF HIS WINE

In front of me as I write is a small menu card, about the size of a postcard, dated October 4, 1973. It reads thus:

PETITS ROUGETS À LA PÉRUVIENNE

CANETONS RÔTIS

PURÉE DE MARRONS

SALADE DE LAITUE

FROMAGE

SOUFFLÉ AUX LIQUEURS

LAFITE 1959

HAUT-BRION 1945

MOUTON-ROTHSCHILD 1920

YQUEM 1945

October 4 was a day like any other day at Château Mouton-Rothschild—any other day when Baron Philippe de Rothschild was in residence. In fact, it was a quiet dinner for six: Baron Philippe and his wife, Baroness Pauline;

a deputy mayor or some such from Bordeaux with his wife; and a couple of Americans who almost never drank '20 Mouton during the week.

The memory of that evening at Mouton and, oh, maybe a half-dozen or so like it over the years came flooding back last week with the news that Baron Philippe had died at his home in Paris at the age of eighty-five. It's a stock phrase, used whenever someone dies, but in his case it is totally accurate to say we shall not see his like again.

There was another evening long after Baroness Pauline's death in 1976, when Baron Philippe got into a heated argument with his close friend and literary collaborator, Joan Littlewood, the English theater producer. At one point, Miss Littlewood left in a huff and didn't reappear.

After about an hour, a butler came in and whispered to the Baron: *"Madame est partie."*

"I know," the Baron replied. "She's probably up in her room."

"No, no," the butler said. "She left. She's gone." Baron Philippe shrugged, and the party went on. Later it turned out that Miss Littlewood had stalked out, ignoring the chauffeurs and cars always available, hiked through the vineyards to a nearby village, where she hopped a bus to Bordeaux, thirty-five miles away, and then took a train to Paris.

Then there is the story about the famous wine writer who used to arrange his visits to Mouton so that they began just as Baron Philippe was leaving for Paris or Denmark or Santa Barbara, California, three of his favorite playgrounds. Each time, as he left, the Baron would instruct his majordomo to see that the writer had everything he wanted.

The writer, purely in the interest of research, of course, would proceed to order up rare vintage after rare vintage, to the dismay of the chateau staff. Finally someone told Baron Philippe. When the writer next visited Mouton, the two men took a stroll in the garden. The Baron asked his friend what he would like to drink for lunch. True to form, the writer mentioned a rare and costly bottle. Throwing an arm around the man's shoulder, Baron Philippe said: "It's not a luncheon wine, old boy, not a luncheon wine." And that was the end of the problem.

The Médoc, the flat, rain-swept peninsula jutting north from Bordeaux between the Atlantic and the Gironde, has changed over the years. Once the vineyard owners sat in their chateaus, arguing over vine rot and genealogy, like the protagonists in a Mauriac novel. Buyers from New York and

Hamburg and Amsterdam would be entertained from time to time, but rarely anyone else.

Baron Philippe changed all that. At any time during the long season that ended after the harvest, his guests might include a famous British actor, a former president of France, a Nobel laureate in poetry, a powerful American publisher, an ambassador or two, a champion automobile racer, an occasional duke, and perhaps even Queen Elizabeth the Queen Mother, who is said to enjoy a decent claret now and then.

Dinner at Mouton was never served in the same place at the same time. It might be a gala, taking up the length of the grand gallery, a room filled with priceless paintings and sculptures. It might be an intimate affair for four at a table in the library. Or it might not be in the chateau at all, but at Petit Mouton, the nineteenth-century house that Philippe de Rothschild lived in, without running water or electricity, when he first came down from Paris to learn the business in 1922.

In preparing her dinners, the Baroness, who had been born Pauline Fairfax Potter in Baltimore and had been a designer for Hattie Carnegie, not only consulted with her chef but also leafed through huge sample books containing pictures of all the chateau's sets of china and table linens. All wines were decanted and served from tall chemical beakers, each with a blue linen cloth tied around its neck. Table decorations were rarely flowers. Mouton employed a full-time horticulturist, who ranged the fields and woods of the Médoc and produced strange Dali-esque table displays from weeds and fungi.

Baroness Pauline was more than a chatelaine. Once at dinner, someone mentioned the English poet Gerard Manley Hopkins but couldn't remember the title of a particular Hopkins poem. Later, in the library, the Baroness spread three books on Hopkins before the astonished guest, one of them opened to the poem he had tried to recall.

Literature was not an affectation at Mouton. Baron Philippe delighted in the company of writers and produced a respectable body of poetry and translations himself. It includes three volumes of poetry; an anthology, in French and English, of the Elizabethan poets; and translations of Christopher Marlowe's *Tamburlaine the Great* and *Dr. Faustus,* and of six plays by the contemporary English writer Christopher Fry, including *The Lady's Not for Burning.*

At Mouton, Baron Philippe's style was Churchillian (he would say

Churchill's style was Rothschildian). He stayed awake talking or reading most of the night, then spent most of the day in his huge bed, which was always littered with pets, books, blueprints, manuscripts, and correspondence. Much of his mail was from hapless Paris bureaucrats whose lives he delighted in making miserable with requests for road closings or road openings and with unrelenting demands that his beloved Mouton be raised to first-growth status.

In 1973, 118 years after the Bordeaux trade rated chateaus Latour, Margaux, Lafite-Rothschild, and Haut-Brion as first growths, passing over Mouton-Rothschild, Baron Philippe achieved his lifelong goal. Mouton was rated a first growth. But it wasn't the bureaucrats who finally relented; it was the owners of the other *premier cru* chateaus and conservatives in the Bordeaux trade who had feared that a change for Mouton would presage change for many other chateaus. These they feared would be demoted or dropped entirely from the list of sixty-two classified growths.

That didn't happen. There was a move to downgrade fifteen chateaus at the time Mouton was elevated, but Baron Philippe, among others, spoke against it. To date, no action has been taken.

Mouton's fate is now in the hands of Baron Philippe's daughter, Baroness Philippine de Rothschild, who has shown every indication of being as devoted to the famed property and its wine as her father.

JANUARY, 1988

## A CRITIC SHAKES UP BORDEAUX

Not since the Bordeaux scandals of the early 1970s, or the 1976 Paris tasting when American wines bested some big Bordeaux, had there been so great a stir in the closed-in little world of fine wine.

The fuss was over an article in the December 2000 issue of *The Atlantic Monthly,* an exceptionally long and mostly favorable profile of the wine critic Robert M. Parker Jr. Copies raced all over the wine world. Winemakers, importers, retailers, and wine stewards talked and e-mailed about little else. With good reason.

Mr. Parker is in this account a rough-around-the-edges John Wayne type, single-handedly taking on the Bordeaux wine trade—definitely the bad guys—by promoting a small group of Bordeaux upstarts who make what are basically American-style wines. Traditional Bordeaux wines are not all that great, anyway, says the author, William Langewiesche, and Mr. Parker is just the man to point out that the emperor has no clothes. At stake, supposedly, is nothing less than the future of the Bordeaux.

"The wine writer Robert Parker Jr. may be the most powerful critic in the world," trumpets *The Atlantic*'s cover blurb. The wine trade, writes Mr. Langewiesche, a correspondent for *The Atlantic* for the last ten years, "has never known such a voice, such a power, before. When it comes to the great wines, those that drive styles and prices for the entire industry, there is hardly another critic now who counts."

Strong stuff. But much of it happens to be true.

Most wine fans, even those who spend big money buying it, actually know very little about it. Through the pages of his bimonthly newsletter, *The Wine Advocate,* which has a circulation of only about forty thousand, Mr. Parker reaches out to that crowd and tells them what they should drink. He makes it easier by rating wines on a hundred-point scale.

Winemakers fear those ratings. Justifiably. Mr. Parker is pitiless with wines he doesn't like. But Mr. Parker's business is finding good wines, and for twenty-two years he has done just that in *The Wine Advocate,* enthusing floridly over several hundred selections in each issue. A formidable task. As in this panegyric for a 1998 Napa Valley zinfandel: "Offering an explosive nose of dried herbs, ground pepper, new saddle leather, rose petals, black raspberries and cherries, this fat, dense, full-bodied luscious wine possesses low acidity, but gobs of fruit."

With no formal training, Mr. Parker has honed his palate to a remarkable degree of sensitivity. In all sincerity he claims to be able to recall every one of the more than two hundred thousand wines he has tasted over the years. Whether he can or not, the fact is that in a world with more than its share of dilettantes and poseurs, he is an unpretentious fellow whose enthusiasm is as unflagging as his capacity for hard work.

Of course, not all winemakers here or abroad, especially in France, are admirers of Mr. Parker. As Mr. Langewiesche accurately notes, many in the French wine trade view Mr. Parker as an interloper, imposing his unsophisticated American taste for wines like the Napa zinfandel on less muscular European wines that rely more on nuance and subtlety than in-your-face aromas and flavors. Nothing wrong with that view. No one at the top of his game—and lots of Bordeaux and Burgundy winemakers see themselves that way—likes being criticized.

But in his article, Mr. Langewiesche has chosen to demonize the French old guard, particularly in Bordeaux, as bitter enemies of progress. Citing the recent phenomenon known as garage wines, a minuscule group of Bordeaux wines much praised by Mr. Parker, Mr. Langewiesche concludes that "the industry worldwide is moving in an unexpected direction, toward denser, darker, and more dramatic wines."

Actually, most garage wines are made as sidelines by owners of traditional chateaus. Sometimes the traditional wines are better than the garage wines. "It would be simplistic," Mr. Langewiesche writes, to believe that the garage wine movement "is entirely due to Parker." But, he says, "these denser, darker wines are the wines that Parker and now much of the world prefer to drink." The Bordeaux wine establishment feels threatened by these new-style wines, Mr. Langewiesche contends, and is "engaged in an increasingly bitter fight against Parker and his influence."

Well, perhaps. Once the Bordeaux wine trade was a closed, secretive world dominated by a handful of powerful exporters, known as *négociants,* who set prices, loaned money, bought wine, and often bought the chateaus themselves. Today many of the old *négociants* are gone. Many of the chateaus are owned by insurance companies, conglomerates, and foreign investors.

Actually the "old families" have been fading away for decades. The collapse of the market in 1974 hastened their demise, humbling some of the most powerful houses, including the Ginestets and the Cruses. If the old-line chateau owners are irritated by the *garagistes,* it is more for the hype surrounding them than for any threat they may pose.

Still, if inky, dark, and as Mr. Parker would say, "explosive" wines are the wave of the future, the Bordelais will eventually get involved. Actually, some of them already have. Over the past decade, comparative tastings have shown that even some of the best-known Bordeaux wines have become softer, richer, and more opulent than in the past. Even ten years ago, distinguishing traditional Bordeaux wines from California cabernet blends was simple; today there are often more similarities than differences.

Who is responsible for that change? Robert Parker, yes. But he's not alone. There will always be hard-liners in Bordeaux bemoaning the passing of the old ways. But there are many more younger people ready to do whatever it takes to make and sell good wine.

The Bordeaux revolution began half a century ago with forward-thinking chateau owners like Philippe de Rothschild and Alexis Lichine, who had traveled the world and were open to modern ideas. Jean-Michel Cazes, the Texas-trained ex-I.B.M. executive who owns Château Lynch-Bages and manages half a dozen or so other properties, has kept the torch burning. At the moment he is arguably the most powerful figure in the Bordeaux trade. Mr. Parker, with his talented palate and his fearless pen, is a worthy addition to this group. Some Bordelais may despise him, but many others are listening to what he has to say. Very carefully.

DECEMBER, 2000

## THE TASTES OF A CENTURY

André Tchelistcheff was a teenager when the Russian Revolution began, living on his family's estate east of Moscow and planning to become a physician. Today, approaching one hundred, he lives in a modest house on the edge of California's wine country.

He never became a physician. He became one of America's, and the world's, most prominent enologists—an honor that, by the way, he has no immediate plans to relinquish.

From 1938 to 1972, Mr. Tchelistcheff (pronounced CHEL-a-chev) was the chief winemaker at Beaulieu Vineyard, where his name became permanently linked with one of America's greatest wines, Georges de Latour Private Reserve. As a consultant to countless other wineries and mentor to several generations of winemakers over more than sixty years, he had a hand in creating the modern California wine industry.

He spends more time in airplanes and on the road than men who are a third his age. His wife, Dorothy, drives him now; he stopped driving in 1990 after he destroyed his Nissan sports car in an accident near here. He emerged virtually unscathed.

Until recently he had thirteen major winery clients in California, Washington State, and Italy. Last week he gave up most of them to return to a job he held fifty-three years ago, working with Beaulieu Vineyards in the Napa Valley, where he will be a consulting enologist.

"I am not looking for a new career, my dear sir," he told a visitor; "I am not going back to compete. But I still have work I want to do." And his remarkable face crinkles into a smile.

Mr. Tchelistcheff is a compact man, barely five feet tall. His manner is courtly, his dress elegant. Even when he relaxes at home, his sweater is draped modishly around his slight shoulders. His voice is heavily accented, easier to understand in French than in English. "Well, when I think of wine, I think in French," he says. His long Slavic face recalls portraits of Nijinsky or perhaps an icon of some forgotten Orthodox saint. Slanted eyes seem to close entirely when he smiles, which is often.

After the Bolshevik victory, he fled to Western Europe. "I saw Communism as a brief cycle in Russian history," he said the other day. "I knew the country would need agricultural specialists one day, so in exile, I studied agronomy."

It was as an agronomist that he settled in France, and it was in France that he became interested in wine and the vines that produce it. Georges de Latour, a native of the Dordogne and the founder of Beaulieu Vineyards, returned to France in 1938 to find a new winemaker. Someone suggested the young Russian émigré André Tchelistcheff, and the rest is California wine history.

Mr. Tchelistcheff is quick to dispel myths about his career at Beaulieu and to credit his mentor for the creation of Georges de Latour Private Reserve, California's first great collectible wine. "I did not create the Private Reserve," he said. "Monsieur de Latour did. I was not the one who first aged the wine in small oak barrels; he was." The first vintage was produced in 1936, two years before Mr. Tchelistcheff's arrival. The winery recently celebrated the wine's 50th anniversary by releasing the 1986 vintage.

Just what a winemaker does is unclear to most wine enthusiasts, but they lionize their favorite winemakers, much as experts on food anoint favorite chefs. Mr. Tchelistcheff detests the phenomenon of winemaker-as-media-star, though he may have been its earliest example. "It means that the winemaker must go out and sell the wine," he said. "And that is not his or her job."

Winemakers are chemists and microbiologists who spend most of their time in laboratories, removed from any glamour in the wine business: the harvest, the crush, the fermentation, and, of course, the parties. In fact, to call Mr. Tchelistcheff an enologist is not entirely accurate, even if he has

spent much of his life in a laboratory. He believes wine is born in the vineyard, which has long been the most neglected part of winemaking.

"We begin in the vineyard," he said, "and, always, we must come back
to the vineyard. Finally, finally, after so many years, the California winemakers are beginning to come back to the soil.

"I think," he continued, "that it was a question of prestige. The winemaker wore a white coat; he was a scientist; he knew the secrets. The
grower was a dirt farmer. What did he know of fine wines?"

Beginning with the 1930s and '40s, when scientists were identifying
particular climates in California, the weather has always been the primary
concern of California winemakers. But in recent years, more and more
winemakers have come to appreciate the European's emphasis on soil and
plant culture. The singular qualities of the B. V. Private Reserve have always
been born in the soil, Mr. Tchelistcheff said—the special soil of the famous
Rutherford Bench region, where the best Beaulieu grapes are grown.

The phylloxera epidemic, which ravaged California vineyards in the
1980s, would never have occurred if the industry had paid more attention
to vine selection twenty years earlier, he believes. "Commercial interests
prevailed," he said. "There are vines that are completely resistant to the
phylloxera, but they don't produce as well. So now the good producers are
being pulled out at enormous cost. It is a real tragedy."

Mr. Tchelistcheff's reputation is founded on cabernet sauvignon.
Beaulieu Vineyards Private Reserve is 100 percent cabernet sauvignon, a
product that he might change now for a more Bordeaux-styled blend.

He may have made great wines from cabernet, the grape of Bordeaux,
but he confesses that his own weakness is for pinot noir, the grape of Burgundy. "I don't drink much wine anymore," he said. "After tasting thirty or
forty wines in a day, I'm too tired. But when we do drink, my wife and I
prefer a good Burgundy."

Which may help to explain why his own favorites among the many wines
he has made over sixty harvests were two pinots noirs, the 1946 and 1947
vintages. He admits he has no idea why those two were so good and subse-
quent ones were not. But now he hopes to find out. "I want to work on
pinot noir," he said. "I would like once again to make a truly fine pinot noir."

As with the 100 percent cabernet, Mr. Tchelistcheff's ideas have changed
over the years. "So much has happened in the laboratory, in the cellar, and

in the vineyard," he said. "I would never take today some of the advice I gave to people fifty years ago.

"For example," he said, "the first Private Reserves, the two made by my predecessor Léon Bonnet, were aged in French oak, barrels Monsieur Latour had bought in France in the 1920s. I changed over to American oak because I wanted more flavor, a more aggressive wine. Today I dislike the American oak and would go back to French."

Mr. Tchelistcheff has always had other clients. Over the years, he did work for Robert Mondavi, Louis M. Martini, Franciscan Vineyards, Conn Creek, Clos Pégase, Swanson, Firestone, Jordan, Buena Vista, Villa Mt. Eden, and Atlas Peak, in California; Château Ste. Michelle in Washington; and Antinori in Italy.

Why, over the long span of his career, was there never a Tchelistcheff winery? "Because I never had the courage," he said, with a touch of irony. "There were plenty of offers. People with money who wanted to be my partners. But remember, I am a child of revolution. I know what it means to lose everything overnight.

"No, I am not a rich man, but I enjoy my life this way: independent."

Few people become rich in the wine business, but some become famous, at least in the industry. Many worked for Mr. Tchelistcheff—among them Theo Rosenbrand, who worked at Sterling Vineyards and is now consulting; Joe Heitz of Heitz Cellars, whose Martha's Vineyard cabernet has long rivaled B. V. Private Reserve; Richard Peterson, who headed Monterey Vineyards and later Atlas Peak; Mary Ann Graf, former winemaker at Simi and now also a consultant; and Jill Davis, the winemaker at Buena Vista.

"There is still another generation of winemakers coming along," Mr. Tchelistcheff said, "and they will lead California in directions we don't even think of now." He predicts that chardonnay and cabernet will fade as the dominant varieties—"I do not believe in the duration of the American market" is the way he puts it—and that there will be more and more interest in other varieties, particularly those of the Rhône Valley in France and of northern Italy that are now grown in California.

"Ah, but to create a great Hermitage—that will take time. It took many centuries in France. And, too, the soils are so specific in the Rhône. I don't think there is any way we will ever duplicate a great Côte Rôtie here.

"Now the Sangiovese, from Italy, that's another story. I have tasted the '86s and '87s in barrel and they are beautiful, beautiful, beautiful!"

But it takes time to create a great wine, and time is money.

André Tchelistcheff is well aware that he will not be around to see many of the changes he predicts. "I don't have much time left," he says, but not in sorrow or anger. He knows he has lived a long life and accomplished most of his goals.

One goal may be beyond his reach: to see his native Russia one last time. "Some friends went," he said. "They told me the estate is gone but that in the graveyard, they found the cross of my grandfather's tomb. Oh, I would like to see that."

Once, years ago, he was invited to come to the Soviet Union as a wine consultant but, as an old White Russian, he was denied a visa. Now he is free to go but he worries about his diet. "Perhaps," he says, smiling. "It may happen. One never knows."

JANUARY, 1991

# EVEN A VISIONARY NEEDS TIME OFF

He was the Peter Pan and Ariel of wine; a man who described one of his own commentaries as "a critique of pure riesling," and who advised on a label that "this wine will accommodate all manner of game and wild beasts, including the sloth."

No, Randall Grahm—for who else could it be—has not left us. But he has gone, as they say in the land of carrot juice and the flapping sandal, to another plane.

Mr. Grahm is the winemaker and wordsmith who in 1993 founded Bonny Doon Vineyard in the hills just north of Santa Cruz. In the years since, he has gained a reputation as one of the country's more imaginative winemakers, a tireless innovator in the cellar and vineyard, and one of the cleverest wine writers.

With Le Cigare Volant, his now famous wine, he set the style for the Rhône Rangers, a group of young winemakers dedicated to the principle that the California wine country resembles the Rhône Valley more than it does Bordeaux and that its wines should reflect this.

Le Cigare Volant, or Flying Cigar, pays homage to Châteauneuf-du-Pape, in the southern Rhône, where in 1954 the town council passed a law forbidding flying saucers, or cigars as they called them, from landing there. Since then, Mr. Grahm points out, none have.

But put all this on hold: There is a new Randall Grahm. Or so it would seem. Gone, for the moment at least, is the insouciance, the lightness of being, the outrageous puns, as in "This brandy to be taken eternally," "One hundred years of Soledad," "Lafite, do your stuff," and "Absinthe makes the Art Garfunkel." Gone, too, are the long parodies in his newsletter of Baudelaire ("Les Fleurs du Malvasia") and T. S. Eliot ("The Love Song of J. Alfred Rootstock"). "The Bureaus of Florid Prose and Adjectiveorrhea have both been indefinitely closed for renovation," is how Mr. Grahm put it.

What now seems evident about him is not self-doubt, but a great deal of introspection and perhaps a slight sulfurous whiff, not of the vinous type but of burnout.

An objective observer would have to say "with good reason." First, there was the loss of his vineyard. Most Bonny Doon fans—and they are legion—are unaware that Pierce's disease, an insect-borne killer, wiped out the main vineyard in 1995. "By the time we realized it was there, it was too late," Mr. Grahm said.

There are other Bonny Doon vineyards near Soledad in Monterey County, southeast of Santa Cruz, where the vines are healthy, and there are grapes that can be bought from vineyards everywhere. There have to be such other routes when you are experimenting with thirty varieties and selling some three dozen different wines. But the original barnlike winery is shuttered most of the time, except for Mr. Grahm's office upstairs, where he often works perched on a large rubber ball ("Good for the spine," he said). But the soul of the venture, the vineyard in the redwoods, is no more.

Then there was the problem with a planned third winery. (His second winery, in downtown Santa Cruz, is in a seedy industrial park populated principally by matte-haired dudes making surfboards.) In 1997 Mr. Grahm announced that he would buy an old winery and vineyard in the Livermore Valley, east of San Francisco. Then doubts set in. At one point he hired a feng shui master, a Dr. Wu, from Los Angeles to check out the suitability of the Livermore site. Dr. Wu said the site was "very good for a graveyard."

But his opinion wasn't the problem. "I fear I may have been done in by my eternal archnemesis, that old commitment thing," Mr. Grahm reflected.

To bolster his own hypothesis, he noted that at forty-five he is still not married, even though he would like to be, or thinks he would like to be.

With the old vineyard and the putative new winery both history, it was inevitable that Mr. Grahm turn his inner light to his—and everyone else's—wines. He has become an almost fanatical exponent of the European principle of *terroir,* the idea that a great wine is the product of soil, climate, and weather, and that the winemaker's duty is to let nature take its course and not foul things up. "In the Old World," he said, "winegrowing is the process of uncovering the expressiveness of the vineyard and the varietal." In this country, he believes, the process has barely begun.

"The wine gestalt is an interplay between surface and depth," he said. "We in this country can polish the surface, but it's a façade; we haven't got the depth. Fortunately, most people like the surface. The marketing people tell us the wines are selling well, but don't we in the small hours have the feeling that we are really putting one over on *tout le monde?*"

To find the depth he feels American wines lack, Mr. Grahm experiments with obscure grape varieties. He long ago abandoned the accepted standards—cabernet sauvignon, merlot, and chardonnay. "We live in a chard- and cabocentric world," he once said.

Instead, he has been a pioneer in working with the Rhône varieties like syrah, grenache, mourvèdre, marsanne, and roussanne. More recently, he has worked with Italian and Spanish varieties, the Portuguese grapes used in Port, and less familiar French grapes like pinot meunier.

Currently he is intrigued by tannat, the grape of Madiran, a wine of southwestern France. "It makes a hard wine in France," he said, "but a rich, soft wine in Uruguay." There is already a Bonny Doon Madiran, but it isn't easy to find.

Mr. Grahm's own future is unclear. Bonny Doon goes on, possibly in a new Santa Cruz winery. He is thinking about a sabbatical of sorts, possibly at the feet of Nicolas Joly at Savennières in the Loire Valley.

Mr. Joly, who makes the white wine known as Coulée de Serrant, is the leading proponent of biodynamics, a system of holistic agriculture developed by the Austrian philosopher Rudolf Steiner. Biodynamism takes organic farming to its extreme. Planting and pruning are timed to the position of the planets, harvest time is determined by the phases of the moon, and vines get homeopathic treatments rather than fertilizer and insecticides.

More and more of France's and Italy's best winemakers use some or all of the biodynamic methods.

Fortunately for us, Mr. Grahm takes enough time out from wrestling with his muse to make good wine. Though his Le Cigare Volant is his best known, his zinfandel, which he named Cardinal Zin, is, in his words, "pure sunshine."

In his line of Italian-style wines, Ca' del Solo, one of them, Il Pescatore, is a delicious white. His Pacific Rim riesling actually includes about 20 percent wine from Germany. As Mr. Grahm notes, "The heart has its rieslings." Domaine des Blagueurs is a syrah from France. Then there are the *eaux de vie* in a variety of flavors, several *vins glaciers,* or California ice wines, and a couple of dynamite grappas.

Suffice it to say that good things emanate from Bonny Doon, and when Mr. Grahm is back in the groove, there'll be more: wines to match or surpass a 1994 pinot noir he called "a bridge over truffled waters," a 1995 sauvignon blanc virtually guaranteed to "enhance your capacity to wonder," and a nonvintage Clos de Gilroy (grenache) "wonderful with grilled *n'importe quoi.*"

NOVEMBER, 1998

## SEEKING HARMONY, FINDING CABERNET

There are pessimists who say the heart has gone out of the wine business, that the pioneers and risk takers have abandoned the stage to number crunchers and marketing types and that one day, not too far off, all wines will taste pretty much the same.

They haven't met Dick Grace.

It is said that there are only three wines in the world as eagerly sought after as the cabernet sauvignon from the Grace Family Vineyards: a Bordeaux, Château Pétrus; and two Burgundies, Romanée-Conti and Le Montrachet, both from the Domaine de la Romanée-Conti.

A spiritual man, Mr. Grace takes his success in stride. Some years ago, in researching his family name, he discovered that one meaning was "an unwarranted gift." And that, he said, "is exactly what the vineyards and winery have been to our family." A former marine, a recovering alcoholic who cannot drink his own wine, a Buddhist and a deeply committed philanthropist, he sees his winery more as an integral part of his spiritual quest than as a business enterprise.

"My family had no money," he said, "and when I started to make some, I had no control." A successful career as a stockbroker and a budding wine business were not enough to hold him together. By the early 1980s, he had a drinking problem. A brush with clinical depression prompted him to take stock of himself.

Today Mr. Grace notes proudly that he hasn't had a drink in fourteen years. He still judges his wines—by bouquet and color.

"When I reviewed my life, I saw it was all about self," he said. "Not that I was exceptionally selfish—it's just that everything I did, every decision I made, was self-generated, self-willed. And I realized that a self-willed life was doomed to failure. We are here for another purpose."

Searching for harmony and peace in his life led him—inevitably, he says—to Buddhism. For each of the last six years, he has spent a month in Nepal, part of it working in one of Mother Teresa's hospitals in Katmandu, part of it trekking to a Buddhist monastery near Mount Everest, where he spends time in prayer and meditation. And the trips to Nepal have fostered a strong friendship between the Graces and Sir Edmund Hillary, the first man to climb Everest.

Through auctions, his wines have raised hundreds of thousands of dollars for charities that assist sick children and their families. Among his proudest possessions are photographs of ill and crippled children that he and his family have helped.

The fame of the tiny Grace winery in Napa Valley far outstrips its minuscule production of ten barrels of wine, or about three thousand bottles, in a good year. Every bottle is sold directly from the winery to a select mailing list of three hundred customers. Each customer is entitled to three bottles and one magnum a year. Should anyone decide not to buy, there are twelve hundred names of people who are waiting to become customers.

Three bottles of the 1994 vintage sold for a total of $275; each magnum cost $395. And everyone pays the same price, including the few restaurants fortunate enough to be on the list, like the Four Seasons in New York. Retailers routinely offer $175 a bottle to people on the Grace list, then sell the wine for as much as $250 a bottle. At commercial wine auctions, Grace wines regularly bring higher prices than even the most famous French wines.

A portion of each vintage, including a few imperials (large bottles containing the equivalent of eight regular bottles of wine), are set aside for charity auctions. At a Birmingham, Alabama, auction, an imperial of the 1993 vintage sold for $18,000, a double magnum for $8,000, and a single magnum for $5,000.

Grace Family Wines have received high praise from the time the first vintage appeared twenty years ago. Only one wine is made, cabernet sauvignon, and it is always 100 percent cabernet sauvignon. The wine has been impeccable, as befits one made from handpicked and severely selected grapes. It is intensely concentrated, deeply colored, beautifully rounded. There are no harsh tannins, even when it is young. How long it will last is anybody's guess—even the earliest vintages are said to be in near-perfect condition.

Dick Grace's favorites are 1980, 1985, 1987, 1990, and 1994.

Is the wine worth the high prices it brings? It is without a doubt the equal of the finest cabernets made in the Napa Valley. At the same time, it is the subject—and beneficiary—of a great deal of hype. The price is a function of the wine's scarcity as much as its quality. It also reflects the effort that goes into etched and hand-painted bottles and handsome wooden cases.

The Grace saga began twenty-five years ago. Then, as now, a stockbroker with Smith Barney in San Francisco, Mr. Grace was looking for a weekend retreat. In a Napa Valley restaurant, he and his wife, Anne, ran into a friend who sold real estate. "He showed us this place," Mr. Grace said, spreading his arms to encompass the Victorian house that is his home. "We bought it on the spot." The Grace family also includes two sons and a daughter.

Originally the Graces planned only to garden, but an acre of cabernet sauvignon vines came with the house, and one day Mr. Grace decided to cultivate them. "I had no idea that the place would one day be so famous," he said, "but I did resolve to make them as good as they could possibly be."

The first harvest was in 1978. Caymus Vineyards, which makes some of the finest cabernets in California, agreed to buy the grapes to blend in with its own. When the Grace grapes arrived, Caymus's owner, Charles Wagner, was so impressed that he decided to vinify them separately. When bottled, the wine had a Caymus label with a special designation: Grace Family Vine-yards.

Through 1983, the Grace wines were made under the Caymus label. In 1985 and 1986, they were made at Caymus and finished in the Graces' cellars. The small Grace winery, just north of the house, was opened in time to handle the 1987 harvest and was expanded in 1992. By then, another acre of grapes had been planted and production had been expanded to include the wines of two other small properties: Hartwell Vineyard, in the Stag's Leap region of the Napa Valley, and 29 Vineyard, a couple of hundred yards south of the Graces on Highway 29.

"They use our grape clone," Mr. Grace said, "and they share our dedication to quality." Hartwell produces twelve barrels a year, and Vineyard 29 about thirty. They share their mailing lists with Grace, and their wines are also in demand. Another name will soon be added: Barbour Vineyards. Jim Barbour was the first vineyard manager at Grace.

The original Grace vineyard was one of the first closely spaced vineyards in the Napa Valley. There were eleven hundred vines an acre, as opposed to the normal five to six hundred. Close spacing stresses the vines; the yield from them is lower, but the grapes are more intense. After the 1994 harvest, this vineyard was pulled up; it had been decimated by a virus known as oak root fungus, and the vines had to be destroyed. The one-acre vineyard was replanted in 1995 with a remarkable 3,465 vines.

"That's six times what is normal in the valley and three times what we planted in 1976," Mr. Grace said.

The second Grace vineyard will have to be replanted next year; it has been attacked by the phylloxera bug. "We've had a couple of tough years," he said. "We got seven barrels in 1994 before we pulled up the vineyard. We got only three barrels, or about nine hundred bottles, in 1995 and about two barrels, or six hundred bottles, this year. I think we'll get one more harvest from the upper vineyard."

Grace customers will probably get a single magnum from the 1996 harvest and perhaps less from 1997's. In 1998, the original vineyard should be back in production.

Then, too, the Grace Family winemaker for seven years, Gary Galleron, left in 1995 for another job. He was replaced by one of the Napa Valley's promising young winemakers, Heidi Peterson Barrett, who is the daughter and wife of prominent winemakers. She is also responsible for other sought-after small-production wines, including Screaming Eagle, Paridigm, and Dalle Valle.

Mr. Grace says he has been offered other projects during what he calls his vineyards' "down time." "Unfortunately," he said, "they were not spiritually in line with what we do. Our goal is to be more than financially successful. We want to be spiritually successful as well."

The Graces' idealistic outlook has never stood in the way of prosperity. Ten years ago, Anne Grace and two friends got together $5,000 and founded the Napa Valley Mustard Company to sell condiments and jellies. In 1995 it grossed $10 million.

No wonder the cork in every bottle of Grace Family Vineyards bears the words: "Be Optimystic."

NOVEMBER, 1996

## SHE PUTS ONE OF THE GREAT VINEYARDS ON PARADE

May-Éliane de Lencquesaing was striding through her new barrel cellar and found the lights weren't working. "They said they'd be here tomorrow," an assistant said.

"Today," she demanded. "I want them here today, and I want this work done."

Mrs. de Lencquesaing (pronounced LONK-sang), seventy-six, is the owner, director, and demanding manager of Château Pichon-Longueville Comtesse de Lalande, one of the great vineyards of France.

This day the pressure was on because Mrs. de Lencquesaing was gearing up for Vinexpo, the wine fair held every other year and scheduled to begin here in mid-June. Thousands of importers, exporters, wine buyers, winemakers, wine writers, and wine enthusiasts would be in town for a week, and like every other famous chateau in the Bordeaux wine country, Pichon-Lalande, as it is informally called, would be host to hundreds of them.

A lucky few would be guests in the chateau. Hundreds would come for lavish dinners. Hundreds of others would turn up to taste wines, visit the cellars, and see the chateau's glassware collection.

Nor was that all. Mrs. de Lencquesaing had bought another chateau,

Bernadotte, a mile or so away, where major renovations were under way to accommodate the overflow of guests from Pichon-Lalande. "It's a charming estate, with one hundred acres of vines, a lovely house, and one bathroom," she said. "We're installing four more."

Bordeaux chateaus are not ordinarily bought casually, but if anyone did so, it would be Mrs. Lencquesaing. By birth, she is a Miailhe (pronounced mee-AYE), a family of wine merchants and growers whose history in the wine business predates the French Revolution and who have owned all of or shares in chateaus Margaux, Ducru-Beaucaillou, Verdignan, Coufran, Citran, Siran, Dauzac, and Palmer.

The family's Bordeaux interests were supplemented by holdings in land and shipping in the Philippines, where some of Mrs. de Lencquesaing's ancestors settled in the seventeenth century. Most of those holdings were confiscated in World War II and have been only partly returned since then.

Mrs. de Lencquesaing's father, Édouard Miailhe, died in 1959—"a great vintage year," she noted—but his estate wasn't divided until 1978. She inherited the family share of Pichon-Lalande, 55 percent, and subsequently acquired the rest.

Originally she had little thought of taking over Pichon-Lalande. In 1950 she had married a French army officer, Herve de Lencquesaing, and they had spent the next twenty-six years at postings around the world, including Fort Riley, Kansas, where to keep busy she went to an army school and became a nurse. Her husband, by then General de Lencquesaing, died in 1990. The de Lencquesaing family owns estates in the north of France, between Arras and St.-Omer in Pas-de-Calais.

For a time in the 1960s, one of Mrs. de Lencquesaing's brothers, William Alain, ran Pichon-Lalande. He quit in a family feud, and from 1972 to 1978 the property was managed by Michel Delon, who owns the nearby Château Léoville-Las-Cases.

In 1855 the Bordeaux wine trade made a classification of the best wines of the region. The top five chateaus were in the first group, the fifteen next best in the second group. Pichon-Lalande was placed in the second group. In recent years, Pichon Lalande and several others in that group have come to be known as "super seconds" because they often match the famous "firsts"—Haut-Brion, Lafite-Rothschild, Latour, Margaux, and Mouton-Rothschild—in quality and, more important, because they can demand nearly comparable prices.

Under Mrs. de Lencquesaing, new cellars have been built, vineyards expanded and replanted, and most of the winemaking equipment has been modernized. But not all the old skills are gone. "We still clarify our wine with egg whites," she said, "six to a barrel, which means six thousand eggs and six thousand leftover yolks."

Many of Pichon-Lalande's achievements in the post–World War II years were the work of the longtime manager, or *régisseur,* Jean-Jacques Godin, who was first hired by Mr. Delon. In 1992, following an internal dispute over questionable winemaking practices, Mr. Godin was dismissed. Thomas Do-Chi-Nam, a young enologist who took charge with the 1993 harvest, is the current winemaker.

There are actually two Château Pichons: Mrs. de Lencquesaing's Pichon-Longueville Comtesse de Lalande and, just across the road, Pichon-Longueville Baron, which is owned by the giant French insurance company AXA.

From the middle of the seventeenth century to the middle of the nineteenth, there was one estate, Pichon-Longueville Baron. Because of inheritance laws, the estate was divided in half in 1850, but it continued to be operated as a single winemaking entity until the 1860s. That was when Countess Marie-Laure-Virginie de Lalande built a separate winemaking site on her side of the road. Her portrait dominates the main salon of Mrs. de Lencquesaing's chateau today.

The descendants of the Baron de Pichon-Longueville sold Pichon Baron to the Boutellier family in 1933. They sold it to AXA in 1987. The decendants of Countess de Lalande sold Pichon-Lalande to Mrs. de Lencquesaing's father and uncle in 1926.

Mrs. de Lencquesaing said she believes that AXA would like to reunite the two Pichons, but she is committed to keeping Pichon-Lalande in the family. She is thinking of passing on the management to one of her granddaughters. She has three daughters, one son, and ten grandchildren.

Referring to Claude Bebear, AXA's president, Mrs. de Lencquesaing said: "At dinners he's proposed toasts to the reunion soon of the two. He's even gone after my children's shares." AXA officials say they have no interest in acquiring Pichon-Lalande.

Besides, what would the 172-acre Pichon-Lalande be without its dynamic owner? Not since the late Alexis Lichine ranged the world on behalf of his Château Prieuré-Lichine has Bordeaux had such a tireless ambassador. One day in Frankfurt, the next in Los Angeles, the day after that in Kobe or Hong Kong or Bangkok, she will readily preside at three tastings of her wines before lunch and another before dinner. Her schedule is likely to include two trips a year to the Nordic countries, three or four to Switzerland and the United States, plus regular visits to Germany, the Netherlands, Belgium, Japan, Taiwan, Singapore and, more recently, to China.

Day-to-day business at the chateau is handled by an office manager. An office for visitors looks after tourists, estimated at six thousand a year, many of whom arrive in tour buses. From April to October, there are two or three dinners a week for up to one hundred guests and frequent lunches with importers, journalists, and winemakers.

When Mrs. de Lencquesaing gives dinners abroad, she often takes copies of the chateau's Limoges dinner service—to be sold to the dinner guests. In the future, she reasons, they will be more inclined to serve Pichon-Lalande wines if they are eating from Pichon-Lalande plates.

If there is a key word in Mrs. de Lencquesaing's vocabulary, it's "discipline." "I was an army wife for twenty-six years, and the army taught me the need for discipline, attention to detail, and the need for effort," she said. "Never give up when you are in trouble. Every day is a battle."

In Bordeaux, Mrs. de Lenquesaing is known as "La Générale." And she doesn't seem to mind at all.

JUNE, 1997

## A BORN SALESMAN

Alexis Lichine died in 1989, at seventy-six, at Château Prieuré-Lichine, his home in Bordeaux. We had come in the end to think of him as a kind of *monument historique,* dominating the wine scene the way the Eiffel Tower dominates Paris. Like the tower, Alexis Lichine was once controversial, even hated, but like the tower, he lived to become a symbol of what is best about France.

He wasn't French, of course, but neither were Picasso, Chagall, Modigliani, or Yves Montand. He was Russian by birth, American by preference, and French in spirit.

And that spirit endured until the end. A week before he died of cancer, battered by painkillers and the ravages of his affliction, he could still critique a glass of wine in a few precise words. And on a brief drive through the vineyards, he could, smiling, point out a chateau whose pompous owner he had insulted with brio many years before.

Being an American gave him the right, or so he thought, to tell everyone from the mayor of Bordeaux to taxi drivers that the only thing wrong with France was the French. That, coupled with his impatience with revered tradition for tradition's sake and his brash approach to the wine trade, did not always endear him to his fellow chateau owners or to the powerful wine shipping firms in Bordeaux and Burgundy.

Today many of the innovations he helped to introduce to the French wine trade fifty years ago are standard practice. Many of the old families survive because their sons have M.B.A. degrees, or the sense to hire people who do. At the end, few members of the fine wine trade in France would deny the enormous influence Alexis Lichine had on their business in the formative postwar years.

His influence extended to relatively minor things. Incredibly, in the early 1950s the vineyards of Bordeaux, including the most famous, still were tilled by horses. He introduced the first tractor ever used in the Médoc in 1954, having purchased it in Burgundy, where, as he once said, "the farmers may look poorer than the Bordelais, but are usually much richer."

As a shipper himself, he traveled tirelessly, promoting and selling his wines when others were content to receive potential buyers in Bordeaux and entertain them lavishly. When most chateaus were strictly private enclaves, opened only to a privileged few guests, Alexis threw open the doors of his beloved Château Prieuré-Lichine to all comers.

There was always a welcome, too, for tourists in the wine country who stopped by unannounced. Billboards along the roadway out of Bordeaux announced—to the horror of the other chateau owners—that Prieuré-Lichine was open daily for tours. Today some of the chateau owners who hated the Lichine billboards are investing as much in visitor facilities as they are in new barrels.

The two Lichine books, his *Guide to the Wines and Vineyards of France* and *The Encyclopedia of Wines and Spirits,* published by Alfred A. Knopf in 1984 and 1967 respectively, were mostly written by others but were checked and revised meticulously by Alexis himself. The books also provided one of his greatest pleasures, which was to range over France, visiting winemakers, restaurants, and hotels.

In Burgundy and Alsace and Beaujolais, where he was a pioneer in urging the small estates to bottle their wines rather than sell them to the cooperatives and shippers, he was hailed as a hero, often by the sons and daughters of men and women he had bought wine from in the 1950s. After the first flutter of surprise, a bottle would be produced, a 1945 or a 1969, and Alexis would talk about the old days. "You weren't even born," he would tell the beaming winemaker, "and your father kept horses where you have that Mercedes now."

The restaurateurs, world-famous now, ran simple country inns when the Lichine visits began. They, too, would pull out wonderful old wines, reminiscing about legendary meals and deploring the perverse modern desire to be thin.

A born salesman, Alexis was also a conscientious winemaker. When he purchased the Prieuré in 1951—it was called Château Prieuré-Cantenac then—the estate consisted of about 32 acres of mostly run-down vines. Today it consists of some 170 acres of first-rate vines, producing on average twenty-five thousand cases of wine a year. In his later years, he doubled the size of the barrel-aging cellars, then doubled them again. Only a year before his death, a handsome new tasting room, which also can be used for large dinner parties, was added to the cellars.

In the tourist season (the Bordeaux coast boasts some of the best beaches in Europe), Prieuré-Lichine's open-door policy requires a staff of five or six, mostly bilingual young British women, who conduct tours through the cellars and sell wine and artifacts, including the two Lichine books in seven or eight languages, T-shirts, corkscrews, and other items more likely to be seen in Napa Valley wineries than in France. From time to time, Alexis would report at dinner that some prominent chateau owner who had loudly deplored Lichine's commercialism had been seen poking around in the gift shop, undoubtedly picking up ideas.

Someone once said of a famous man that his greatest creation was himself. Alexis Lichine came about as close as anyone does to being truly *sui generis*. The petty details of his life could be vague: the early days in New York and Paris, the army, chats with Churchill and Eisenhower, the friendship with Grace Kelly, real-estate deals in the Caribbean, paper mills in Angola. But what are details when one is fashioning a legend?

There were setbacks: Alexis Lichine & Company was frittered away, and with it the professional use of his name; Château Lascombes, purchased soon after the Prieuré, sold; portions he owned of two great Burgundy vineyards, Mazis-Chambertin and Bonnes Mares, sold. He would have loved to have crowned his career by running, if not owning, a first-growth chateau—Haut-Brion, perhaps, or Margaux—but that was not to be.

Nor, for all his prestige, was he ever one of the inner circle of chateau owners, most of them tradition-bound conservatives who resented his aggressiveness and self-promotion. He was too independent, too much a star, too much a loner. He was hurt when groups of chateau owners made promotional tours to the United States and Britain and failed to invite him. "He would have taken over from the start," an owner once told me. "He was larger than life; he could never conform to the will of the group. It would have become his trip."

Bordeaux, to Alexis, was France in microcosm. He loved the gentle, almost flat countryside, the orderly, cosseted rows of vines, the cool pine forests, and the vast sky, but the people drove him to distraction. Each edition of his guide has carried a swipe at the city of Bordeaux for its obstinate refusal to mark the routes from midtown to the wine country. He alternately mocked and raged at what he considered the secrecy, the snobbery, and the petty jealousies that for so many years prevented the winegrowers from opening themselves and their chateaus to a growing wine public. "They are too busy hating each other, spying on each other," he would whisper dramatically, sounding like John Barrymore as Richard III.

For all that, he had heroes, mentors. The first was Raymond Baudouin, founder of the *Revue des Vins de France,* who taught him how to taste and buy in a hundred musty cellars, and how to eat and drink in the golden days of Alexandre Dumaine at La Côte d'Or in Saulieu and Fernand Point at Pyramide in Vienne. There was Seymour Weller, who ran Château Haut-Brion for his cousins, the Dillons, and who guided Alexis during his first tentative days in Bordeaux. Ronald Barton at chateaus Léoville-Barton and Langoa-Barton, and later his nephew Anthony Barton, were friends and supporters, as were the Marquis de Lur-Saluce at Château d'Yquem and Hugues Lawton, from the old Bordeaux shipping family.

Then there was Philippe de Rothschild, at Château Mouton-Rothschild, whose consummate taste, flamboyant lifestyle, and seemingly bottomless pocketbook Alexis so admired. The Médoc is an immense open place and there was enough room for two such towering egos, but only barely. What they shared, usually at arm's length, was the conviction that Bordeaux had to be opened to the world. With frequent caustic asides, each respected the other for his contributions to that cause.

In a *New Yorker* profile some forty years ago, Joe Wechsberg called Alexis "The Pope of Wine." "Missionary" would have been a better analogy, though not from one of the mendicant orders. "He eats, sleeps and thinks wine," Wechsberg wrote, and so he did. Imperious, querulous, egotistical, he managed to convey his enthusiasm and to garner more converts to wine single-handedly than the rest of the trade—grown to an industry now—and its legions of drumbeaters and press agents had ever managed.

He was one of a kind, and we shall not see his like again.

JUNE, 1989

## JOURNEYING THROUGH SPAIN INTO THE FUTURE

We had been driving all day, had tasted a lot of wines, and were, frankly, starved. So we watched impatiently as Manolo Perez grilled forty tiny lamb chops over vine cuttings.

We were in the small cookhouse off the dining room at the Bodegas Hermanos Perez Pascuas, one of the best-known wineries in Ribera del Duero, in northern Spain. When we finally sat down, we paired the succulent chops with some of the bodega's best wines, several vintages of its Vina Pedrosa, one of the most sought-after wines in Ribera del Duero.

The dinner and the wines were special because they showed how far Ribera del Duero has come. When I first visited northern Spain, almost twenty years ago, it would never have occurred to me to venture outside of Rioja, then the country's only serious area for the production of red wines. Today Ribera del Duero is one of Spain's fastest-growing wine regions.

In the United States, Spanish wines were once found almost exclusively in Spanish restaurants. Now they are suddenly popular everywhere. American wine fans are discovering regions like Navarre, Priorato, Toros, and Ribera del Duero that were virtually unknown a few years ago. Even Rioja, the best known of all the Spanish wine districts, is back in fashion. And it still dwarfs the others: All of Ribera del Duero annually produces about 1.5 million cases of wine, while some of the larger Rioja bodegas have more than 2 million cases aging in their cellars.

The region of Ribera del Duero is about two hours north of Madrid in the province of Castile and Léon. It stretches sixty-eight miles along the banks of the Duero River between Soria in the east and Valladolid in the west.

Grapes apparently were cultivated and wines made in the Ribera del Duero by the Romans, but modern plantings date from the Middle Ages, when French monks brought vine cuttings to the Cistercian monastery at Valbuena de Duero, about twenty miles west of Pedrosa. Ribera del Duero is not a monoculture region; corn, vegetables, and sugar beets are grown here, along with the sheep that provided us with the lamb chops at Perez Pascuas. Fifty years ago, vines were pulled out to plant the more profitable beets. Now new vineyards are seen everywhere.

Paradoxically, while the Ribera del Duero was dismissed as a serious viticultural region as recently as the 1980s, it has been home to Spain's most famous winery, Vega Sicilia, since 1846. Vega Sicilia wines, which are blends of local grapes and Bordeaux varieties, developed an almost legendary reputation for their scarcity, quality, and staggeringly high prices, a reputation they hold even today—but with lots of competition.

Originally the Ribera del Duero was a region of small unskilled winemakers. Cooperatives were set up in the 1950s, but their wine was not much better.

No longer. In a tasting of some three dozen Ribera wines put on for us by the Consejo Regulador, the regional governing body for wines, there were at least twenty that I would rate as very good to outstanding. In addition to the bodegas discussed here, I was particularly impressed by the wines of the Bodegas Valduero in Aranda, and not just because the winemaker, Margarita Madrigal, has such a lovely name. Like all the better Ribera wines, the Valdueros have depth and intensity and show a restrained use of oak. If there is any fault common to many of the Ribera del Duero wines, it is the vanilla taste imparted by the excessive use of new oak.

The man who changed everything was Alejandro Fernandez, who had made a fortune designing machines to harvest sugar beets. As a boy he had made wine with his father, and in the 1970s he realized his dream—a bodega of his own at Pesquera. Inspired by Vega Sicilia, he determined to make wine in his own style, using only local grapes, especially tinto fino, which is virtually identical to the tempranillo grape of the Rioja.

Mr. Fernandez now operates two huge bodegas in the region to keep up with the demand for his Pesquera wines. We had lunch with him and his daughters, Eva Maria, a trained enologist, and Lucia, who handles the business side, one rainy afternoon at Condado de Haza, the newest Fernandez bodega, in Roa de Duero.

Warmed by a crackling fire and a succession of local dishes, we tasted through—drank, actually—ten or twelve wines ranging from his regular Pequeras to Janus, a famous *gran reserva,* a sweet red wine best forgotten, and a heavy Port-like red from Toro, a relatively new wine district west of Valladolid.

Spanish laws allow for three categories of fine red wine: *crianza, reserva,* and *gran reserva. Crianza* is a wine released for sale in its third year after having spent at least six months in barrel (in Rioja, twelve months); *reservas* must spend three years in barrel and bottle; *gran reservas* must spend two years in barrel and three in bottle or vice versa, and cannot leave the bodega until the sixth year after the vintage.

Mr. Fernandez's original enologist was another legendary figure in the region, Teofilo Reyes, creator of Protos, a wine almost as famous as Vega Sicilia but, until recent years, only within Spain. Mr. Reyes now makes wine under his own name at his own bodega in Penafiel, in the warmer western part of the region, near Vega Sicilia, Pesquera, and Protos.

Like the Napa Valley in California, Ribera del Duero has begun to attract outside investors who have succumbed to the romance of wine—Francisco Rodero, the fashion designer from Barcelona, for example. Mr. Rodero, who was born in Pedrosa, owns the Bodegas Rodero, just down the road from the brothers Perez Pascuas. Rodero wines are sold in the United States under the Carmelo Rodero label. (The Perez Pascuas, by the way, are veterans by Ribera del Duero standards. Their bodega dates from 1980 but they were making wine at the same site long before that. They grow all their own grapes, many of them on forty-year-old vines.)

Several of the best bodegas in Ribera del Duero are actually just outside the legal boundaries of the appellation. One afternoon we dropped in on Mariano Garcia Fernandez at Bodegas Mauro in the village of Tudela de Duero, which is almost in the suburbs of Valladolid. Once the winemaker at Vega Sicilia, Mr. Garcia now focuses on his own wines, which are rich and concentrated but not as powerful as the Vega Sicilia wines. He said he would like one day to rival his former employer with Terreus, an impressive *gran reserva.*

Also beyond the viticultural pale is the huge Abadia Retuerta at Sardon, east of Tudela, a former monastery converted into a winery by Novartis, the giant Swiss drug manufacturer, which has hired Pascal Delbeck, long associated with Château Ausone in Bordeaux, as winemaker. Like Vega Sicilia, Abadia Retuerta uses Bordeaux varieties in some of its wines. Its 1997 is a blend of 65 percent tempranillo, 30 percent cabernet sauvignon, and 5 percent merlot.

I had been told that visiting Vega Sicilia would be a memorable experience, and it was. The immense new winery looks as though it might have been designed by Frank Gehry, whose Guggenheim Museum in Bilbao is just two hours to the north. At Vega Sicilia, stainless steel, brick, rare woods, superb lighting, and extraordinary artwork have produced a building only a few other wineries in the world could match.

We tasted through all the Vega Sicilia wines, from the simplest, Tinto Valbuena, through the rare Unico, which usually contains about 20 percent cabernet sauvignon and normally spends about ten years maturing, first in wooden vats, later in barrels, and finally in bottles. I've tried Unicos before. Typically, the one in barrel at the bodega was an exceptionally big, rich wine. Whether it could ever be paired with food is questionable. It doesn't seem as if it would want to share the spotlight with anything.

At a smaller Vega Sicilia winery nearby we tried Alion, a new wine the bodega is developing, as well as samples of a variety of experimental wines. Some of them included wine from Toro, which is being promoted as another trendy wine area and produces big rustic wines resembling zinfandels from the Sierra foothills in California.

One winemaker I didn't get to see in Ribera del Duero, but would have liked to, is Peter Sisseck, a Danish-born and Bordeaux-trained enologist who works for several bodegas but also produces his own wines in a small bodega in the riverside town of Quintanilla. His first vintage of the wine Dominio de Pingus (only a few cases were made) sells for about $500 a bottle when it can be found. Rave reviews from several wine critics, including Robert M. Parker Jr., whetted the appetites of collectors around the world. When a ship carrying most of the wine to the United States broke up at sea, the wine was lost and the price of what remained shot up.

I had a chance to sample the wine at a restaurant in Bilbao. Like the Vega Sicilia and a few California cult wines, it is tremendously rich and concentrated and seems to overpower anything paired with it. Not that most purchasers will have that problem. Five-hundred-dollar wines are trophy wines, meant to be shown off, not drunk.

MAY, 2000

## NEVER DRINK AN INFERIOR BOTTLE

How old are you, mate?" Len Evans asks.

Taken aback, the other fellow mumbles something unintelligible.

"I'd say you're about about sixty," Mr. Evans says, "and from the looks of you"—he pokes at the man's ample waistline—"you'll be lucky to make seventy-five. You've got about fifteen years ahead of you, and it's time for you to learn my Theory of Capacity."

Leonard Paul Evans is indisputably Australia's best-known wine figure. He is a winery owner, a wine writer, and a wine judge. He made wine drinking acceptable in a macho beer-drinking country, and he has inspired a generation of younger Australian wine lovers. He is a true hedonist and, oh yes, a philosopher.

"You've got to make the most of the time you've got left, man," he says, philosophically. "You've got to calculate your future capacity. A bottle of wine a day is 365 bottles a year. Which means you've probably only got 5,000 bottles ahead of you. People who say you can't drink good stuff all the time are fools. You *must* drink good stuff all the time. Every bottle of inferior wine you drink is like smashing a superior bottle against a wall; the pleasure is lost forever. You can't get that bottle back." A proper sense of urgency established, he opens a bottle of Bollinger 1985.

Mr. Evans practices what he preaches. His cellar is vast and his generosity is legendary. His longtime friend and colleague James Halliday said, "No one has ever been permitted to drink a mediocre bottle in Len's company."

Like so many other people involved in wine, Len Evans is where he is mostly by chance. A Welshman by birth, he had hoped to become a professional golfer in England but tired of teaching duffers. In 1953 he emigrated to New Zealand and worked as a forester. Moving to Australia in 1955, he was at one time or another a laborer, a welder, a forester again, the manager of a duck farm, and on weekends, a serious golfer. He also found time to write, submitting humorous scripts for radio and television shows. He even took a job washing glasses in a Sydney restaurant.

From lowly glass washer he moved up to become assistant manager of Sydney's biggest hotel, the Chevron Hilton. Those were the years, the 1960s, when he developed his love for wine. He founded the Friday Table at the Hilton, where wine enthusiasts would meet weekly to eat, drink, and teach each other about wine.

In 1969, on borrowed money, he opened his own restaurant, Bulletin Place, in central Sydney. With tasting rooms and a retail wine shop, Bulletin Place became Australia's unofficial wine capital. "It was 'Taste this,' 'Try that,' 'Open another bottle' from morning to night," Mr. Evans recalls. "Those were heady days."

All during this period he was writing about wine. His *Complete Book of Australian Wine* has been through five editions, the latest in 1990.

In 1969 he joined with other investors to start the Rothbury Estate, a winery in the Hunter Valley, the best wine region of New South Wales, some three hours' drive north of Sydney.

The 1970s, when Australians first discovered wine, were for Len Evans what his Australian biographer, Jeremy Oliver, calls the golden years. Backed by a wealthy friend, Peter Fox, he founded the Evans Wine Company. With Rothbury well under way, they acquired two chateaus in Bordeaux: Rahoul, in the Graves region, and Padouen in nearby Barsac. They agreed to buy Alexis Lichine & Company, which included the famous Château Lascombes in Margaux. With help from Robert Mondavi, they bought a vineyard in the Napa Valley in California. And they invested in Petaluma, a prominent winery near Adelaide in South Australia. The future seemed limitless.

Then, in December 1981, Peter Fox was killed when his Ferrari missed a turn on a road near Sydney. He was forty-three. He was also, it turned out, in financial difficulties. Soon, Rahoul, Padouen, and the Napa Valley vineyard were gone, and the deal for Lascombes could not be closed.

Whatever effect the crumbling of the empire had on Mr. Evans—and friends said it was devastating—he has survived. The Evans Wine Company has been reborn as the modest Evans Family Winery, with headquarters at the gate to Mr. Evans's 350-acre estate, Loggerheads. The wines are made at Rothbury nearby and sold mostly through the mail.

Now, when he isn't sculpting stone monoliths, a recent hobby, or working in ceramics or driving golf balls or judging wines or raising millions for charities or imparting his own brand of wine enthusiasm to audiences all over Australia and the United States, Mr. Evans pretends he is ready to take up the life of a country squire at Loggerheads.

And of course, there is the Theory of Capacity. At seventy, he still has quite a few bottles to open. "And open them I will, mate," he says, "beginning right now, with another Bollinger."

MARCH, 1993

## GROWER'S COMMUTE: SCARSDALE TO FRANCE

When Stephen Schecter looks out over his vineyards, he recalls his childhood in Queens, shakes his head in disbelief, and says, "When my father talked about red or white, he meant horseradish."

Mr. Schecter is the chatelain of Château Pech de Jammes, a seventeenth-century wine estate at Vayrols, in the hills southwest of Cahors in south-central France. He is a member of the Confrérie du Vin de Cahors, an association of prominent producers, and he is a well-known figure at the local supermarket, where the locals are not yet accustomed to Americans coming by to load up the station wagon.

In 1985 Mr. Schecter was a successful New York investment banker with a wife, two children, and a more or less traditional nine-to-five, Scarsdale-to-Wall-Street routine. Today he is all that and something more; he is also a winegrower in Cahors.

Over the years, many Americans have invested in wine properties in France, and some even work the land themselves. But few have become a part of the wine community so quickly as Steve Schecter and his family. Mr. Schecter is a managing director of Wertheim Schroder & Company, an investment banking house in New York and London. In his early days at the company, Mr. Schecter shared an office with Roger Yaseen, for many years local doyen of the Châine des Rôtisseurs, a worldwide gourmet society, and an incorrigible wine collector. A few years across the desk from Mr. Yaseen, and Mr. Schecter was hooked. He and his wife, Sherry, began touring wine regions and building their own wine collection.

They also dreamed of becoming winemakers themselves. Wertheim's merger with Schroder, a London house, in 1986 provided Mr. Schecter with some unexpected but serious money and an excuse to spend time each month in the London office. Already Francophiles, the Schecters began upgrading their winemaking dream. Why not, they asked each other, a chateau of our own?

A magazine advertisement led them to the real-estate experts at Christie's, the London auction house. "They had four places," Mr. Schecter said. "Three in Bordeaux, where we assumed we really wanted to be. Two were too run-down to be considered; the third didn't even make wine."

Mrs. Schecter recalled: "The man told us about the fourth place, but without much conviction. It was in Cahors, two hours from Bordeaux. He didn't think we wanted to go that far. But I saw that look in Stephen's eye and I knew we'd be up at five the next morning and on the road. We were."

Some 150 miles east of Bordeaux and 70 miles north of Toulouse, Cahors is the capital of the Quercy region and is deep in what President François Mitterrand liked to call *la France profonde,* "the real France." Cahors is one of the oldest viticultural regions in the country; its wines were known to Horace and Virgil. In the Middle Ages, Cahors wines were used to strengthen the wines of Bordeaux, then thin and pale, and make them fit for export. So rich were the Cahors wines that they were known as *vins noirs,* "black wines."

Cahors wines went into a decline in the fifteenth century. They came back briefly in the nineteenth century, just before all of France's vineyards were devastated by the phylloxera epidemic. So severe was the disease that the region did not recover until after World War II. But it was not until 1971 that Cahors gained the coveted *appellation controlée* status for its vineyards.

The man credited with reviving the fortunes of the Cahors wine trade is Georges Vigouroux, a wine merchant who, beginning in the 1970s, restored the estate of Haute-Serre on a stony hilltop not far from the Schecters' property. He cleared trees, dynamited boulders, and broke up subsoil that had not been worked in a hundred years. He planted 150 acres of the local auxerrois grape—called *malbec* in Bordeaux—built a winery, and slowly began to make the wines of Cahors prominent once again.

When the Schecters first fell in love with Pech de Jammes in 1986, they could see Haute-Serre off to the southwest, little realizing it would soon be central to their new life. They bought Pech de Jammes from Bernard Pons, secretary general of the neo-Gaullist party in France and a prominent figure in the governments of Georges Pompidou and Valéry Giscard d'Estaing. It was Mr. Pons who, as minister of agriculture in 1971, was able to have Cahors elevated to A.O.C., or Appellation d'Origine Contrôlée, status, instantly increasing the value of the vineyards and making some of his grower friends overnight millionaires.

He also asked the University of Toulouse to undertake a study of the terrain at Pech de Jammes, which he had owned since 1961 and which was a working, if not overly productive, vineyard. Under the university's tutelage, forty thousand vines were raised in hothouses while the old vineyards were cleared to receive them. Replanting began in 1973 and the chateau received its own A.O.C. rating in 1978.

Early on, Mr. Pons had Mr. Vigouroux take control of his vineyards in exchange for a share of the profits from their wines. Haute-Serre crews tended the vines, picked the grapes, made the wine, and sold it under the Pech de Jammes label. When the Schecters arrived in 1987, they found this arrangement still in place and asked for some changes.

They planned to do their own marketing, for one thing; they wanted new bottles and new labels; and they wanted to have some say on the vinification and aging of their wine, even if it was being done at Haute-Serre. There was considerable tension at first; Mr. Vigouroux and his team did not know what to make of this energetic couple from New York and their new ideas.

"Eventually we sat down and talked it out," Mrs. Schecter said, "and now we're really good friends." Mrs. Schecter, who once worked in financial public relations in New York, grew up in North Bergen, New Jersey but went to school for a time in Switzerland and speaks excellent French. With it, and with a warm, outgoing personality, she soon made a wide circle of friends among the conservative winegrowers and their families.

"It's remarkable how quickly the Schecters became part of the community," Mr. Vigouroux said. He in turn spearheaded efforts to induct Mr. Schecter (in France the name is pronounced shesh-TAIR) into the Confrérie du Vin de Cahors, making him the only American member.

The Schecters divide their time between Scarsdale, New York, and Cahors, with stopovers in London. To do it—they are in Cahors at least a week every month—they spend enormous amounts of time in the air. Door to door, with a plane connection in Paris to Toulouse and an hour's drive from Toulouse to Cahors, the trip takes about fifteen hours.

"It is a long trip," Mr. Schecter said during a lunch by his swimming pool, "but we're so eager to be here, we really don't mind it at all."

The first buildings at Pech de Jammes were erected in the 1770s and the most recent in 1794. Since buying it, the Schecters have restored the old walls, discovering in the process windows and passageways that had been blocked off. One unused wing has been converted into quarters for their two sons, Larry and Scott, and a small barn is being made into a guest house.

Of the forty-three acres on the estate, twenty-five are in vines. Plans call for doubling that through leasing planting rights and purchase of more land. But there is no timetable. "You learn right away that nothing moves swiftly," Mr. Schecter said. "If you go next door to borrow a hammer, it's bad form even to mention the hammer for the first hour."

OCTOBER, 1989

## A CELLAR MASTER OF
## A CERTAIN VINTAGE

It's easy to get tired of New York in midwinter: the crowds, the angry drivers, the dirty snow. Some people head for Miami. Me, I take the A train to Brooklyn. With the sun on the water, the Brooklyn Bridge soaring overhead, and Lower Manhattan a thousand yards away, there is no place like it to see New York. It's a John Sloan painting and an Edna St. Vincent Millay poem, and an afternoon spent taking it in will get you through to April better than a weekend in Paris.

In fact, I had a second agenda when I walked down Old Fulton Street last week from the subway stop to the River Café, the restaurant nearly under the bridge. Someone had told me that the average time a California winemaker stays at any one job is four years. With new restaurants opening almost hourly, sommeliers in California and everywhere else seem to move every six weeks. I've heard that "Excellent choice, sir" line from wine stewards I'd swear were not old enough to drink two years ago. Joe Delissio has been the wine director at the River Café for twenty-three years. That has to be some sort of record, and I was curious to see how he was holding up.

Mr. Delissio started out as one of those "excellent choice" kids: He was twenty-three when Buzzy O'Keeffe, who owns the River Café, hired him as wine director. He has literally grown up with the wine boom. "My father owned a saloon in Brooklyn," said Mr. Delissio, who is now forty-six, "but I knew nothing about wine, nothing. It was what the guys drank in my father's place when they were off the hard stuff."

Mr. O'Keeffe didn't know much about wine either, and the Frenchman who had been buying wine for him had his own way of adding up bills. "I was working as a psychiatric therapist during the day and an aspiring rock musician at night," Mr. Delissio recalled. "Buzzy hired me because my mother, who did his books, vouched for me. Actually, it wasn't that hard. I bought the wine and I sold it. It's like Pete Rose used to say when someone would ask him how he did what he did—'I see the ball,' he'd say, 'and I hit the ball.'"

It was more complicated than he admits, starting as he did from zero. "At first I listened to the salesmen," he said, "even if I didn't like the wine. I figured I didn't know anything. Eventually I decided that it was time to buy what I liked. I traveled to California and to Europe. And it was a lot easier after that."

Mr. Delissio no longer hovers over nervous guests, recommending Volnay with the veal. He chooses the wines, buys them, coddles them. He consults with the chef, Brad Steelman, to match wines with special dishes, like a Cune Reserve 1991 with the buffalo steak.

Mr. Delissio has worked with many River Café chefs, including Larry Forgione and Charlie Palmer. These days he trains others to push the Volnay. Right now it is Scott Calvert, who has been with the café for seven years but sommelier for only a few weeks.

Have tastes changed? "When I started, about a third of our liquor business was wine," Mr. Delissio said. "The rest was spirits. Now, 75 percent of our alcohol business is wine. In the '60s our wine sales were almost all French. So was our wine list. Then I got fascinated by American wines, long before they were really popular. And now we seem to be going back to French wines."

The Delissio theory: It's all economics. When French wine prices go through the roof, American wines become the favorites. When their prices balloon, the French come back. Wines are not inexpensive at the River Café, but they can be less costly than at other places around town, particularly at the top end of the price list.

Like Harry's at Hanover Square, Veritas on East 20th Street, and other restaurants that offer hard-to-find wines at decent prices, the River Café has a policy of buying and holding rare bottles. Thus Mr. Delissio can offer seven vintages of Château Pétrus, six of Château Rayas, a rare Châteauneuf-du-Pape, and nine vintages of Le Montrachet from Domaine Ramonet. The average price for a bottle of wine is around $55.

The roughly five-hundred-bottle list is not the longest in the city, but it is one of the broadest. Not many restaurants in New York offer California cabernets going back to 1969; few if any offer white Burgundy verticals from Ramonet and Leflaive, along with red Burgundy verticals from Leroy and Romanée-Conti. A Château Lafite-Rothschild 1982 goes for $1,000, but some restaurants charge three times that.

The large-bottle collection is exceptional as well. How about, for example, an imperial—six liters—of Beaulieu Vineyards 1991 Georges de Latour Private Reserve? Or a double magnum of the 1974 Robert Mondavi Reserve Cabernet Sauvignon? The cost is $600 and $1,200, respectively, but these bottles are not for ordinary mortals.

The café has an unexpectedly good selection of Spanish wines and an unexpectedly short selection of Italian offerings. Port, Madeira, and Cognac are well represented. The Sherries include a fino flown in fresh from Spain.

"In the old days, people asked for recommendations," he said. "Now all they want to know is if it's from a hot new winery and does it have a high score from the critics.

"Suggest a wine with some age? Forget it. Everyone wants the most recent release from some fancy little winery that has no wine for sale, anyway. It's a dangerous trend."

Recently Mr. Delissio began to oversee the wine operations at Mr. O'Keeffe's other restaurants, the Water Club and Pershing Square. And he has formed a consulting business to aid places whose sommeliers don't have twenty-three years' experience. There was a book last year, *The River Café Wine Primer* (Little, Brown), and he writes about wine for *Bartender* magazine.

With all this, I hope he doesn't leave the River Café. It's comforting, dealing with a pro.

JANUARY, 2001

## POL ROGER'S AMBASSADOR

Every wine region of France has its dynamic female executives. Philippine de Rothschild at Château Mouton-Rothschild in Bordeaux, is a
classic contemporary example. But the Champagne country has been producing exceptional female leaders for almost two centuries.

Odette Pol-Roger was perhaps the last of these Grande Dames of
Champagne. A popular figure in Paris society and a lifelong friend of Sir
Winston Churchill, she died at eighty-nine last Christmas day at her home
in Épernay, in the Champagne district.

The first major figure among these women, and still the best known,
was Nicole-Barbe Clicquot-Ponsardin, who took over the Clicquot
Champagne house at the age of twenty-seven, following the death of her
husband. She opened her own bank in Rheims when she was forty-four
and remained active in the Champagne business until shortly before her
death in 1866, also at eighty-nine. The widow Clicquot has become im-
mortalized as Veuve Clicquot, the label seen on the house's prestigious
Champagnes.

Mathilde Émile Laurent-Perrier ran Laurent-Perrier from 1858, after
her husband's death, until 1890. The firm was later purchased by another
Champagne widow, Marie-Louise Landon de Nonancourt, whose son
Bernard brought it into prominence in the years after World War II. Louis
Pommery ran Pommery & Greno from 1858 to 1890, and, more recently,
Camille Olry-Roederer took over Champagne Louis Roederer in 1932 on
the death of her husband. She died in 1974. Lily de Lauriston Bollinger
took the helm of Bollinger on her husband's death in 1941 and remained
in the business until she died in 1977.

Odette Pol-Roger never formally ran Pol Roger, but she was a director
and played an active role in its management. After Winston Churchill
placed a wreath at the Tomb of the Unknown Soldier in Paris on November 11, 1944, he dined at the newly reopened British Embassy. The ambassador, Duff Cooper, and his wife, Diana, seated the prime minister beside
Mme. Pol-Roger. He was so charmed by her that he told his hosts to invite
her to dinner whenever he was in France. Later that month, for his birthday, she sent him a case of his favorite Champagne, Pol Roger 1928.

For the rest of Churchill's life, she sent him the 1928 Pol Roger on his birthday, until it ran out in 1953, and thereafter the 1934. She was invited to dine at Downing Street whenever she was in London. In 1952, Churchill named one of his racehorses Pol Roger and brought Mme. Pol-Roger to England to watch the filly run. It eventually won four races, including the Black Prince Stakes Handicap on the day of Queen Elizabeth's coronation in 1953. Churchill never visited Mme. Pol-Roger's home at 44 Avenue de Champagne in Épernay, but he was fond of describing it as "the world's most drinkable address."

When Churchill died in 1965, Mme. Pol-Roger was one of the personal friends asked to the funeral. The Pol-Roger family paid its respects by bordering in black all the bottles of its Champagne exported to Britain. In 1984, the Pol Roger company named its finest bottling Cuvée Winston Churchill, reserving the initial release for the British royal family and Churchill's descendants.

Susan Mary Alsop, widow of the columnist Joseph Alsop, described the relationship between her friend Odette and Churchill to Cynthia Parzych and John Turner, authors of *Pol Roger,* a history of the Champagne house. It was, she said, "a beautiful December–May relationship, quite harmless, and smiled upon by Mrs. Churchill." For Churchill, she said, "Mme. Pol-Roger personified France, and he was a romantic about France."

Mme. Pol-Roger was the great-granddaughter of Sir Richard Wallace, a member of Parliament and a connoisseur whose vast art collection was left to the British government in 1897. Years later, when Odette Wallace and her two sisters, Jacqueline and Nicole, were reigning young beauties in Paris, they too were known as "the Wallace collection."

Odette met Jacques Pol-Roger, the grandson of the founder of the Champagne house, and later one of its partners, in Épernay while her father, a French Army general, was stationed there. She married in 1933 and soon became involved in the family business. "When I was younger," she told an interviewer in later life, "the smart Champagnes in Paris were Pommery and Clicquot. I set about getting Pol Roger in there." She discreetly urged friends in Paris society to serve Pol Roger at their daughters' weddings.

Her firm was founded in 1849 by Pol Roger, Pol being another spelling for Paul. When he died in 1899, his sons Maurice and Georges, changed their surname to Pol-Roger, using the hyphen. During World War II, the Pol-Rogers, like many Champagne families, used their agricultural supplies to care for their workers and soldiers wounded in battle. Several times, Mme. Pol-Roger bicycled to Paris, a twelve-hour trip, to carry messages for the Resistance. Once she was arrested in Paris and briefly detained by Gestapo agents for wearing a Royal Air Force pin on her lapel.

After her husband's death in 1964, she continued as a director of Pol Roger while devoting time to gardening, decorating her homes in Épernay, Paris, and Normandy, and to her favorite pasttime, fly-fishing. She recalled once leaving a dull Paris dinner party and deciding to drive straight through to her home on the little Andelle river in Normandy. "It was daylight when I arrived," she said. "I was about to go to bed when I looked out the window and saw a lovely trout jumping in the stream. Still in my dinner clothes, I ran out with my rod and caught him."

FEBRUARY, 2001

## THE IMPRESARIO

The end of a special era in the American wine trade drew ever closer recently with the death of Alexander C. McNally. Mr. McNally, who was sixty-two, had been living quietly in retirement in Hartford, largely unknown to a younger generation of wine drinkers and collectors. From 1969 to 1984, however, he was one of the most visible figures in the American wine trade. It was during those years that he ran the annual Heublein Premiere National Auction of Rare Wines.

At a time when most Americans considered wine exotic and vaguely suspect, Mr. McNally, who was known as Sandy, employed his considerable knowledge and flair for showmanship to break down their reserve—and part more than a few of them from their cash. Then as now, Heublein Inc. owned Beaulieu Vineyard, a prominent California winery, and was a major importer of fine wines. Heublein is now a subsidiary of Grand Metropolitan P.L.C., a British conglomerate. The auctions moved from city to city each year, in classic show-business style. They were held in hotel ballrooms and similar venues and cleverly promoted by advance tastings in other cities around the country, usually with Mr. McNally presiding. This way, each auction garnered weeks of national and local news coverage for Heublein—and for Mr. McNally.

Sandy McNally was a Princeton man born in Indiana. Like his fellow Midwesterners, F. Scott Fitzgerald, another Princetonian, and Cole Porter, who went to Yale, Mr. McNally adopted East Coast sensibilities along with an accent and demeanor that hinted of old money, a bit of which he had, and a Continental background, which he did not have. Not entirely, anyway.

He did spend five years in Bordeaux, working for the Russian-born wine figure, Alexis Lichine. It was an apprenticeship in wine and worldliness second to none. Most of Mr. Lichine's postulants lasted less than a year, mentally and physically drained by his czarist temperament. Mr. McNally luckily arrived with plenty of hubris of his own. He had founded a wine club at Princeton before he was old enough to drink and loved to recall how he had bluffed a Princeton, New Jersey, wineshop into opening an account for him, an unknown and underage student, by striding in and demanding: "Tell me, my good man, what do you have in a decent Pommard?"

America has had knowledgeable wine collectors since the days of Washington and Jefferson. Prohibition rendered all alcohol, even fine wine, slightly louche. The Heublein auctions helped make wine respectable again and, at the same time, democratized it. People with money discovered that buying an old Lafite-Rothschild not only was fun and a good investment, but also got you on television.

Mr. McNally orchestrated his tastings with the flair of a Diaghilev. To wield the gavel, he brought J. Michael Broadbent, the dean of wine auctioneers, from Christie's in London. To hold each bottle aloft, as Mr. Broadbent extolled its virtues, he hired Roy Sheppard, an actor, and dressed him in powdered wig, monocle, satin knee breeches and buckled shoes. Mr. McNally shared the podium with Mr. Broadbent and Mr. Sheppard, but he was always ready to talk to reporters and hold up a bottle or frown into a glass for the cameras.

Today a wine auction held by Christie's or Sotheby's would be judged a failure if it generated less than $5 million in sales. In their first six years, the Heublein auctions took in a total of less than $1.3 million. But with no precedent in this country, the sales were considered quite successful.

Even with inflation, the prices they brought would reduce a contemporary collector to tears. In 1971, for example, the sale included case lots of Inglenook Napa Valley cabernet sauvignon from 1887, 1892, 1897, and 1933 through 1936. The 1887 fetched $5,000 but the 1935 and 1936 sold for $710 a case. In 1972, the 1929 Château Cheval Blanc was knocked down for $350 a case. Two years later, Mr. McNally was happy to sell the entire five-thousand-bottle cellar of Claude Philippe, the legendary wine director of the Waldorf-Astoria Hotel, for $80,000.

As the Heublein auctions moved into the 1980s, wine retailers became worried. Auctions, they concluded, could siphon off a sizable chunk of their business, even when, as was the case, laws required that the auction sales be cleared, for a fee, through local retailers.

Faced with mounting legislative opposition generated by the liquor trade, the Heublein auctions ended in 1984. Wine auctions returned to New York in 1992, but only in forced collaborations between retailers and the auction houses. Contemporary auctions have occasional dramatic moments, but the truly rare old wines are gone, and so is the McNally touch.

Collectors often prefer to sell their wines anonymously. But insiders smiled when they spotted a rare bottle in the Heublein catalog described as being "from the cellar of a Scottish gentleman." To themselves, they added: "From Indiana."

JULY, 1997

# CHAPTER 6

# WINE, RESTAURANTS, AND (SOMETIMES INTIMIDATING) SOMMELIERS

It's easy to get the impression from wine books that it's advisable to train like a fighter going into the ring before coming up against a wine steward and an intimidating wine list. It's difficult but not that difficult. Some of the secrets are here in this chapter, including one that reveals most wine stewards are really not ogres out to humiliate you. Ordering wine is actually rather enjoyable once you get the hang of it.

Sending back a bottle that you think is bad is a bit more difficult but—guess what?—some of the best sommeliers in the country are worried because they think not enough people do it. The staggeringly high prices of wine in recent years can make anyone nervous, especially someone who had planned to spend $20 and finds the cheapest bottle on the list is $45. Some restaurants offer better prices than others; it's worth seeking them out.

Daniel Johnnes is a sommelier's sommelier as you will discover even if he never worked in Harry's New York Bar in Paris. There is no longer a Harry but, as one of these essays notes, his successors are there to greet you at *"Sank roo doo noo."*

## IF THE WINE IS OFF . . .

Everything up until now has been amateur night. All the training, all the sweat was nothing but preparation for this moment. It's Randolph Scott climbing out of the back cockpit and saying: "Take her up, kid. She's all yours."

You're about to send back your first bottle of wine. Don't expect much help. No one can go with you into this no-man's-land. Your guests don't know Lambrusco from Léoville-Poyferré. And your wife thinks it's all nonsense anyway.

The waiter will think you are trying to impress your guests. And the wine steward, if it's that kind of a place, may take it as a personal affront. But you think the wine is bad and you're going to say so. So do. At best you will meet your trial with steadfastness and courage, facing down your adversary like El Cordobés. At worst you can console yourself that you've come a long way from the days when you were afraid to order the bottle in the first place.

Have you ever been out with people who ordered wine, then made no fuss when it turned out to be bad? Rest assured, there is a lot of bad wine consumed by people who actually know it is poor but are afraid to say anything.

There is a way to do it, of course. It is a little ploy that seeks to involve the waiter in the painful process of rejection. Once you've made your fateful decision, call him over and say: "I think this wine is off." Don't use the word "bad." It's an angry word and gets people angry. On the other hand, "off" sounds slightly technical, so he might think you know something. Then quickly add, "What do you think?"

If you're lucky, he will agree—only because he doesn't know much more than you do. If he calls over the maître d'hôtel, you are in deeper water but you still have some momentum on your side. In the best of all possible worlds both of them will bow deferentially, saying, "Of course sir, forgive us. We will replace it immediately."

If they stonewall it and declare, "Seems okay to us," well, there is nothing you can do but get tough. One gambit, albeit a dangerous one, is to announce loftily, "I think it's off and I want it replaced, even if I have to pay for it."

No place will let you pay for the first bottle, no matter how many dark looks they may direct your way. Some years ago, after three days of tasting several hundred red wines at the Los Angeles County Fair, I drove north to San Francisco where, among other places, I lunched at a dramatic but awful restaurant in the Embarcadero Center. I was served a perfectly rotten zinfandel from a highly respected winery. It was just bad.

The young woman took it away. Ten minutes later she returned to say, "The manager is a wine expert and he says the bottle is good. You don't have to drink it but you have to pay for it." I allowed as how I had yet another choice: to walk out. Which I did.

Of course there is another side to this coin. There are people who really do send bottles back to impress their guests, and there are people who send bottles back simply because they don't know what a good Bordeaux tastes like. These people are a problem for restaurateurs. The late Henri Soulé of Le Pavillon had a technique for the showoffs who announced their 1959 Lafite was no good. First the captain tasted the wine, then the sommelier, then the maître d'hôtel, then the great Soule himself. By then everyone in the restaurant was watching and the Hollywood mogul or captain of industry was squirming in his seat. Inevitably all the staff would agree that the wine was perfect. Just as inevitably, Mr. Soule would agree to replace it, of course. It cost him money but it tamed or got rid of another obnoxious guest.

A guest who thinks a bottle is bad because he or she simply doesn't know the taste of Bordeaux or Hermitage or Barolo is usually handled in much the same way, but less theatrically. Again, a good place will offer something else, counting on having made a friend.

The ultimate solution is, of course, to lay the whole problem on the restaurant. This is what I am eating: What should I drink with it? Let the restaurant decide.

But suppose it's still a bad bottle? Then what? Then refer to paragraph five above. Or switch to beer.

OCTOBER, 1981

## THE RETURNS ARE IN

Someone mentions sending back wine at a restaurant, and I think of a fearsome editor to whom I once was indentured. One day, a young reporter returned to the office without the interview he'd been sent to get.

"I got thrown out," said the terrified cub.

"You get back there," roared the boss, "and tell them I won't stand for that kind of treatment."

Sending back wine is like that. It's all very well for me to tell you not to take any guff from a surly waiter or a patronizing sommelier. But being there, screwing up the courage to say "I think there's something wrong with this wine," that's something else again.

"He jests at scars," Romeo said, "that never felt a wound."

The truth is, most of us simply don't have the guts to do it. Wine intimidates us—particularly the rituals that surround it. We're already in over our heads, we tell ourselves, and it will be simpler just not to get involved. Even people who know their wines hesitate to complain. They know that many restaurant people can't recognize a bad bottle of wine when they taste (or smell) it. Everyone who regularly drinks wine in restaurants has encountered the waiter—and even the sommelier—who says, icily, "There's nothing wrong with this."

The most common problem in bottled wine is corkiness. The smell of a corked wine is reminiscent of a damp basement; it's slightly sour and slightly moldy. Frank Schoonmaker, in his encyclopedia, likened the smell of a corky—or corked—wine to that of rotten wood. He said that there was nothing to do with such wines but pour them down the drain, which is right. He also said corkiness happens to only one bottle in a thousand, which is wrong.

Corked bottles are much more common than that—I'd say more like one bottle in every hundred or so. Sometimes the contamination is slight, and you live with it. But if the wine is really dank, it should be rejected. It's not harmful, just very unpleasant.

Every restaurant that is serious about wine should make sure that all its dining-room staff gets to know the smell of corked wine. A corked bottle is not a reflection on the restaurant; it happens, sooner or later, to the very best wines in the very best places, and there is nothing anyone can do about it. Above all, there is no reason for the client to feel guilty; the restaurant invariably passes the offending bottle back to the distributor. He's the one who gets stuck.

Sometimes the problem is not immediately evident. The waiter presents the bottle; you nod knowingly. He opens it, pours a bit, and you taste. Without thinking, you nod again and he pours into everyone's glass. Then, after a minute or two, you say: "Wait a minute; something's not right here. I think this stuff is corked." And you call back the waiter.

"Why didn't you say something before I poured everyone else's?" he asks.

"It took me a while to realize what was wrong," you say, lamely.

Again, it's not your fault. Sometimes a wine is so bad that one whiff is all that's needed. More often, it takes a few minutes to realize that something is wrong. There is only one way to go: Stick to your guns.

The best course, if you think of it, is to enlist the waiter or sommelier's advice. Rather than announcing "Waiter, this wine is corked," say something such as "Waiter, try a sip of this and tell me what you think. It took me a minute, but I'm pretty sure this wine is corked." It may work or it may not, but most people would rather be consulted than told off.

Sometimes a bottle of wine is oxidized. This simply means that air got at the wine in some way and aged it prematurely. This is more prevalent with old wines, but it can happen with fairly young ones. Usually the wine has turned brown and tastes dried out, like dead leaves. It's much easier to detect in white wines, of course, but you will recognize it in red wines as well. In this case, there is rarely any argument from the waiter.

From the restaurateur's perspective, the principal problem is the guest who doesn't know wines very well and who is disappointed by what is, in fact, a perfectly good wine. Anyone, or almost anyone, can enjoy Beaujolais or a robust California jug wine on first acquaintance. A great Bordeaux is something else. The mere ability to pay for it doesn't imply the ability to appreciate it. Fine wines simply are not for everybody; they are, the very best of them, acquired tastes.

Every first-class restaurant has experienced the moneyed boor who hopes to impress by sending back the Lafite. They tell tales of the late Henri Soulé taking wine back at his famous Le Pavillon and then, at the end of the evening, tearing up the entire check and suggesting that henceforth the client dine elsewhere.

Some restaurants have a no-return policy on very expensive wines. Others make a point of discussing the wines with the guest before selling them. The restaurant that stocks exceptionally good—and expensive—wines almost always has a staff that knows how to sell and serve them.

And in fact, ordering wine in a restaurant should be a cooperative venture. Only a clip joint will try to pass overpriced wines on the unwary. If you eat in that kind of place, you deserve whatever happens to you. A serious restaurant has people who look forward to helping you pick a bottle you will enjoy. Ask for help. You will probably pay less than you had anticipated. What's more, you will have established a rapport with the staff that will come in very handy if and when the dread moment arrives when you have to clear your throat and say, "I, uh, er, well, I think there may be something—you know—wrong with this wine."

APRIL, 1987

## BOOZE AND MUSES

Westbrook Pegler dropped into Harry's New York Bar in Paris one afternoon in 1936 and decided that the joint's best days were over. "The old saloon," he wrote, "smells of dead cigars, beer and the past." Peg had a way with words, but it seems not to have occurred to him that at Harry's they sold the past; booze was just a sideline.

Scott and Zelda and Ernest were Left Bank people. Today you'd be hard put to find a waiter at the Dôme or the Select or the Deux Magots who'd ever heard of any of them. But they're enshrined at Harry's—along with Black Jack Pershing, the Dempsey-Tunney fight, Prohibition, raccoon coats, the '29 Crash, the Depression, and the liberation of Paris. They're only yesterday at 5 Rue Daunou.

American or pseudo-American restaurants seem to open every week in Paris. Californian, or Tex-Mex, or "New Yorkais," they work hard at being trendy and seem to find favor with Parisians.

"We're not part of that world," says Andy MacElhone, Harry's son and heir. "We don't serve regular food and we don't sell wine. This is a saloon." In fact, there is food, of sorts: sandwiches at the bar at lunch and hot dogs all day from an antique steamer left over from the 1933 World's Fair. And

Champagne is wine, isn't it? But Andy is right. Harry's is a saloon—a saloon curiously suspended in time.

The years between the wars were good years for Americans here, and not just for the ones striking literary poses in St-Germain-des-Prés. Paris was dirt cheap and drinking was legal and Americans flocked here by the boatload. A dollar was worth 25 francs; a cocktail at Harry's was 4 francs, or about 15 cents. And it was made with real stuff.

Harry's was very much a newspaper hangout. Even during the Depression, some twenty American papers had representatives here, including a black newspaper, *The Chicago Defender.* The reporters congregated at the Hotel Scribe. Harry's, a block away, was social center, watering hole, and spiritual home. Harry bought them drinks, carried their tabs, and heard out their laments. As a closetful of crumbling clippings attests, they repaid him with endless citations, uniformly flattering, in whatever dispatches they managed to send home.

In the 1920s and '30s, Harry's had what amounted to a resident laureate, Sparrow Robertson. Sparrow had been a track star of some note in New York in the 1880s and had come to France with the Y.M.C.A. during the First World War. He hooked on as a sportswriter with the Paris edition of *The Herald Tribune,* now *The International Herald Tribune,* and drifted into gossip writing with Harry's as his base. Day after day, Sparrow chronicled the antics of the local sporting crowd and of the fighters, jockeys, bootleggers, and Hollywood stars who dashed straightaway from the boat train to Harry's.

The newsmen may be gone, but there are always tourists, here and at the twenty or so bars around the world patterned after the Parisian original. Many just walk in, guidebooks in hand. The Ritz, around the corner in the Place Vendôme, sends over quite a few, as do most of the hotels frequented by Americans. The French come too, usually the young who are fascinated by America and who think Harry's is authentic Americana.

But is it? Harry MacElhone, a Scotsman, had worked all over Europe before he came to Paris, but he had never been to America. His son Andy married an American but says, flatly, "I'm a Parisian." Andy's son, Duncan, has a degree from Georgetown and spent a few years on Wall Street, but he's back in the Rue Daunou now. "I love the States," he says, "but I feel that this is where I belong."

The place had been called the New York Bar and was owned by an ex-jockey named Todd Sloan. Harry, hired on in 1911, quit two years later, went to the States, and wound up at New York's Plaza Hotel. He served in the British navy, then worked at Ciro's in London. In 1923 he heard that Sloan's, just across the street from Ciro's in Paris, was for sale. He signed the papers on February 8, 1923, the day Andy was born.

During World War II, Harry returned to London. He was one of the first civilians back in Paris in 1944, retired in 1953, and died five years later. "For years," Andy says, "we'd tell customers, 'He's in the back room,' or 'He'll be back later.'"

The drinks here are 46 francs now, almost $8, but who's going to put a price on an authentic bit of the good old days? The late Simone Signoret, who didn't hang out in Harry's, called her autobiography *Nostalgia Isn't What It Used to Be.* It is at Harry's. (Tell the driver *"Sank Roo Doo Noo."*)

OCTOBER, 1988

## IN RESTAURANTS, STICKER SHOCK

Pity, if you can, the wine lover with a taste for the most expensive wines available in New York restaurants. Prices for these top-quality wines differ so greatly from restaurant to restaurant that a customer keeping one eye on the budget could go into the wine-list equivalent of sticker shock.

Take, for example, one of the most sought-after California boutique wines, the Maya cabernet from tiny Dalle Valle Vineyards in the Napa Valley, which makes only a few hundred cases of it each year. Our well-heeled wine fan would pay $600 for a bottle of the 1992 Maya at Lespinasse, the four-star restaurant in the St. Regis Hotel, but only $200 at the River Café in Brooklyn. Wouldn't you go over the bridge to save nearly $400 on a bottle of wine?

But then, things are not much better for drinkers of more moderately priced wines. A 1992 Gallo cabernet sauvignon from Sonoma County, for example, ranges from $52 at the Oyster Bar to $105 at Barolo.

So what are wine lovers and alert consumers, no matter how deep their pockets, to do? There is no way to determine how or where to get the best prices for wines in restaurants, or to gauge which restaurants consistently offer the best deals.

The publishers of the Zagat survey of New York restaurants have been bothered by this for a number of years, too, and they decided to make a price comparison. They sent out 150 requests for wine lists and from the responses chose ninety-nine lists from restaurants in Manhattan, Brooklyn, and Queens. The results, said Tim Zagat, who publishes the survey with his wife, Nina, will be seen in a guide listing the best—and worst—wine prices, along with price comparisons for specific wines.

Mr. Zagat said he had made a second request to several restaurants that did not answer but that are known to have good wine lists. Specific wines were cited only if they could be found in at least five of the restaurants that responded.

The guide is to be published, and will be available on Zagat's Web site, within six months, but here is a preview: The ten restaurants with the best prices are Harry's at Hanover Square, Chiam, Veritas, Sparks Steak House, Union Square Café, Windows on the World, Tomasso's, Henry's End, Piccola Venezia, and the Oyster Bar.

The ten with the highest prices are Lespinasse, Le Cirque 2000, La Grenouille, Daniel, the "21" Club, San Pietro, La Caravelle, the Four Seasons, Le Bernardin, and Park Avenue Café.

The top-ten and bottom-ten lists were compiled by looking at the number of wines for which each restaurant has the lowest or the highest prices in the city. Harry's offered eighty-five wines at the lowest prices of any restaurant in the survey and none at the highest. Chiam has sixty of the best prices but also had one of the most expensive wines, the 1994 Harlan Estate cabernet sauvignon, at $400.

At the other end, Lespinasse had eighty-five of the highest prices and none of the lowest. Le Cirque 2000, like many restaurants, had wines in each category: it had sixty-eight of the highest-priced wines but also offered twelve wines at the lowest prices in the city.

The wine-by-wine comparisons are startling, as they were in 1998 when Zagat first surveyed restaurant wine prices; that survey was available only on its Web site. The range is most breathtaking among the most expensive wines. Lespinasse listed the 1982 Château Pétrus at $4,500, Le Cirque 2000 had it for $4,000, and Harry's for $1,750. The best price— $800—was at a nightclub called Upstairs at Decade.

Prices, of course, are driven by demand, and the fact that twelve restaurants offered the 1982 Pétrus at $1,750 or more shows the almost mindless demand for outstanding vintages. Château Latour 1982 was listed at $1,900 at La Grenouille, and Harry's was charging $575. But the 1983 Latour, also a great wine, could be had at Harry's for $155.

So how do restaurants set their wine prices? Historically, the restaurant markup has been two and a half to three times the retail price. Some restaurants still hold to that formula, but many have abandoned it.

"I don't follow any rule anymore," said Daniel Johnnes, the wine director at Montrachet. "I used to buy a bottle of wine for $10 and sell it for $30. Now it costs me $20, so I sell it for $40. The profit is the same but the markup has changed. The accountants don't like it, but I feel I don't have any choice."

Restaurateurs cite their replacement costs as the main reason for the staggering prices they charge for certain wines. And in fact, a buyer like Ralph Hersom at Le Cirque 2000 is going to have to pay staggering prices to replenish his small stock of Pétrus 1982. Little of this wine was made, and very few bottles of it ever appear on the auction market. Some restaurants can offer lower prices on this kind of wine because they have no plans to replace it once it is sold.

Replacement costs are becoming a problem for less expensive wines, too. A good review can double or treble the prices of some Bordeaux, Burgundies, and California cabernets that are in short supply to begin with. At the same time, there are many wines in ample supply whose costs are still unconscionably high, simply because of the demand.

Bear in mind that the wine you pay $12 for in a retail shop and see on a restaurant list for $36 probably cost the restaurant $7. A good way to determine the wholesale cost of a bottle of wine is to note what it costs by the glass. Many restaurants sell a glass of wine for what they paid for the bottle. Thus, if there are five glasses of wine in a bottle, the last four are all profit for the restaurant.

Most restaurateurs shrug off complaints about wine prices, pointing to the retail market, where buyers seem willing to pay almost any price for rare wines and even for many wines of much lesser quality. The consumer's only weapon is to search for restaurants that work to hold prices down and to seek out wines that are good values. The Zagat wine survey, when it arrives, could be a great help in this effort.

MARCH, 1999

# THE LONG AND SHORT
# OF WINE LISTS

There are strong men who would rather face a firing squad than a wine list. Men keen of eye and firm of grip, masters of the world they survey, find their vision blurred and their hands atremble as they struggle with the foreign names and, quite often, appalling prices.

The date, or the family, or the colleague stares impassively, waiting, not yet aware that a proud man's self-esteem is about to self-destruct. The wine steward, or captain, or whichever rat it was who supplied the cursed list, has disappeared. What to do, what to do?

The right thing, the sensible thing, would be to call the waiter, any waiter, and say: "Look, I have no time to go through this thing. Bring us something that's good—and cheap."

Waiters respect that sort of plea, even in the best restaurants. They know the wine list is mostly for show, and besides, they now have a chance to move another bottle or two of the Beaujolais the boss told them to push.

Is there an ideal wine list? That's a rhetorical question. The answer is no, there isn't. Which doesn't prevent us from creating one.

A fan magazine called *The Wine Spectator* each year dispenses awards to restaurants with formidable wine cellars. It is a competition where big is better and biggest is best. Now, it's conceivable that a restaurant with one hundred thousand bottles of wine in the basement will have a modest little wine list. Conceivable, but not likely.

For people who know their wine, an outsize wine list is like a seed catalogue to an armchair gardener. The exotic names convey images of the Côte d'Or in autumn, the green Tuscan hills, or the morning mist in the Napa Valley. They bring to mind forgotten dinners and unforgettable wines. It requires a "The man is waiting, dear" to bring the reader back to reality.

There is definitely a place for the comprehensive list. But should the uninitiated have to suffer for the sake of the buffs? Absolutely not. Some restaurants solve the problem by having two lists: the main one and an abbreviated version. Lutèce, perhaps New York's best-known French restaurant, has long offered complete and abridged lists; so has Barbetta's, an elegant Italian restaurant on the West Side of Manhattan. Some restaurants have, in effect, three lists: a long one, a short one, and a half-dozen or so suggestions on the menu itself.

Let's face it, most people order a single bottle of wine. A choice of a dozen or so wines should be more than adequate. A choice of two hundred is overwhelming.

So our hypothetical list is really two lists: an exhaustive one for the serious fan, a shorter version for the faint of heart.

But even the short list, if it's just a list, doesn't address the key problem of what goes with what. Here is where we separate the creative restaurateur from the pack. He or she makes the simple effort of suggesting wines with each entrée on the menu. A sample from such a menu would look something like this:

B R O I L E D   D O U B L E - R I B B E D   L A M B   C H O P S ,   $ 2 3 .

*Lamb demands a robust red. We suggest 1987 Coltibuono Rosso, from Italy, $15; 1988 Beaulieu Vineyard Rutherford Cabernet Sauvignon, from California, $22; or 1986 Château Lascombes, from Bordeaux, $38.*

The prices are approximate and will vary from restaurant to restaurant. The point is to give the customer an intelligent choice, one he can exercise without embarrassment. It's impossible to go wrong with any of the aforementioned wines; choosing among them in this way becomes an adventure rather than a burden.

Some restaurants assume all or most of the decision-making responsibilities. A number of France's most celebrated restaurants, including the three-star Lucas-Carton in Paris, offer *menus de dégustation,* or tasting menus, that pair appropriate wines with an array of specialties from the kitchen. The trouble with these affairs is overkill; they are the gastronomic equivalent of the three-ring circus, with too many different foods—sometimes seven or eight small portions—and too many different wines. One tends to leave the table sated but a little vague about what really happened.

The Cellar in the Sky, a restaurant within the Windows on the World complex atop the World Trade Center in New York, has been orchestrating food and wine for its dinners since it opened in 1976. The dinners offer five courses, and there is a separate wine featured with each course.

Wines are poured generously, so guests can concentrate on the ones they prefer.

Informal restaurants have developed many ways to, as they like to say, "market" wine. Little table tents, most often supplied by a distributor, plug certain wines and usually explain how best to use them. There are "wines of the week" and "wines of the month" promotions, Spanish "fiestas" highlighting Rioja wines, and "Oktoberfests" featuring German or Alsatian wines. They are all good ways to make ordering wine easier in public places.

Then, of course, there is the human factor: the sommelier and the captains who are up on their wines. For some reason, Americans have never been comfortable with sommeliers. It has something to do with being independent. Like when we drive around aimlessly for an hour rather than ask for directions.

The ideal wine list, then, is one that lays it all out simply, so that our hero can order his lamb chops and announce to one and all: "I think I'd prefer the Lascombes '86." Just as if he had known all along.

MAY, 1991

## ON THE CASE

To describe Daniel Johnnes as the sommelier at Montrachet, the popular restaurant in New York's TriBeCa, would be like describing Stephen Sondheim as a piano player. There's a bit more to it.

Mr. Johnnes (pronounced JON-iss) is one of a new breed of wine stewards who is perfectly happy to suggest something to drink with your sea urchin custard but whose job description goes way beyond recommending a white for the fish.

Restaurateurs have always been ambivalent about the wine steward. Some of the world's best restaurants have none. At the Four Seasons, for example, the captains handle the wine chores. At Taillevent in Paris, the owner, Jean-Claude Vrinat, rules benignly—and single-handedly—over one of France's finest cellars.

Historically, the sommelier played a mostly passive role. He was there to look after the cellar and help clients choose wines. If he sold a few extra bottles, so much the better, but pushing the product was not his principal responsibility.

Often the sommelier's most important task was covering for a patron who knew little about wine. Restaurant owners usually came up through the kitchen, a tough world of long hours, low pay, and unremitting work. Not a milieu in which to learn the finer points of wine.

To a degree, that's still true. A busy owner often will leave the wine buying, storing, and selling to a specialist. A specialist like Daniel Johnnes.

When Drew Nieporent, who had been a maître d'hôtel at La Grenouille, opened Montrachet in 1985, it was a true gamble. *Haute cuisine* had yet to become part of the TriBeCa scene. But the inspired cooking of David Bouley was an instant hit. Seven weeks after the place opened, *The New York Times* awarded it three stars.

"It was unbelievable," said Johnnes, in recalling that heady and chaotic time. "We went from 40 covers a night to 150, then to 200. I'd been hired as a waiter, but suddenly Drew needed someone to handle the wine. And I was off and running."

Johnnes handles all matters pertaining to wine for three other restaurants in which Nieporent is involved: TriBeCa Grill, Della Femina, in East Hampton, Long Island, and a soon-to-open establishment, also in East Hampton. "I have the buying power of about $1 million a year," Johnnes said.

In a way, he's prepared for this job much of his life. He was born on the Upper West Side of Manhattan and raised in Scarsdale. His father, a filmmaker, and his mother, a writer, had lived in France for a time, and when his father died in the early 1970s, his mother decided to take Daniel and his sister back there for a year.

They moved to Le Barroux, a village in Provence, with only two hundred people, "a lot of whom spoke only Provençal," he said. Johnnes studied French and fell in love with a village girl. When the family returned to New York, Johnnes spent the next four years virtually commuting from New York to Provence. "One summer," he said, "I worked with a local guy in his vineyard. Another year, I was a waiter at a nudist camp, and yes, I worked in the nude." (The romance did not endure the test of time.)

After graduating from the State University at New Paltz and a brief stint at an office job, he went to work at the Landmark Tavern on 11th Avenue, acknowledging that "the nine-to-five routine was not for me."

Soon he was back in France, apprenticing first with Guy Savoy, a prominent Paris restaurant owner, then with Jean-Luc Poujauran, a Paris baker famous for his sourdough baguettes. Next came a six-month stage at the Relais de la Haute Lande, near Mont-de-Marsan in southwest France. "We did great regional things," he said, "foie gras, confit, tourtières. I still do foie gras for my friends at Christmas."

Back in New York, Johnnes made an important decision: "I was going back to the front of the house. The kitchen was dirty, the hours were brutal, and the pay wasn't great, either." He found a job at Le Régence, the restaurant in the Hotel Plaza Athenée, where he first met Nieporent, whom he would later follow to Montrachet.

Because the restaurant is named for what many connoisseurs believe to be one of the world's greatest white wines, Le Montrachet, from Burgundy, Johnnes has concentrated on building a solid Burgundy list—a labor of love, since he is an unabashed Burgundy fan. "First we had to put together a decent list of Montrachets," he said. Not an easy task, given the price of these rare wines. Le Montrachet and the other, hyphenated Montrachets—Puligny-Montrachet, Chevalier-Montrachet, and Bâtard-Montrachet—are now well represented, but their prices can be formidable. To balance his list, Johnnes has discovered a variety of lesser Burgundies, little-known wines from southwest France, and some intriguing American wines. (At TriBeCa Grill there are fewer superstars in the cellar, but the prices are even more attractive.)

At regular intervals, Johnnes puts on a variety of special dinners both at Montrachet and TriBeCa Grill. Some feature a group of Burgundies; some match the wines of a specific producer. Recently, he also established the Burgundy Club at Montrachet, where members are invited to tastings and dinners built around their favorite wines.

Finally, Daniel Johnnes has become an importer. Daniel Johnnes Selections are wines he picks up during regular trips to France. They are featured at the restaurants and sold in as many retail shops as he can induce to carry them.

Not the image of an old-time sommelier bringing your Pouilly-Fuissé up from the basement, is it? Definitely not, Johnnes insists. "The modern sommelier," he says, "is an educator and an ambassador who must also define the image of his restaurant."

Well said—so long as the Pouilly-Fuissé is properly chilled.

FEBRUARY, 1993

# A CONNOISSEUR OF WINE, FOOD, PEOPLE

Louis P. Daniel's death in October 1998 at age seventy-seven did not really end an era in New York restaurant life; that happened years before. But it did bring back a host of memories, especially for wine lovers.

For nearly two decades, Mr. Daniel owned Le Chambertin, a French restaurant at 348 West 46th Street. To its regulars, Le Chambertin was a special place. Mr. Daniel, known to all of them by his first name, was a hands-on restaurateur, constantly moving from the front door to the bar to the kitchen to the customers' tables. Few of his customers were famous, but he knew many by name.

It was not by accident that Mr. Daniel, a Breton, named his restaurant after one of Burgundy's most famous vineyards. He was passionate about wine. He knew every bottle in the converted coal bin that was Le Chambertin's cellar, and long before computers, he knew what he sold and exactly what he wanted to buy. He could and would terrorize hapless wine salesmen who knew less than he thought they should.

This was unusual, because French restaurateurs were rarely wine enthusiasts. Even famous restaurants often depended on wine salesmen to choose the wines, keep their cellars stocked, and put together their wine lists. Some still do.

Not only did I learn about wines from Mr. Daniel, but I learned how they were sold. Before it was fashionable, he offered wine dinners. Looking through his cellar, he would find wines that had reached their peak, then offer them as part of special dinners. I remember one in the early 1970s that included a bottle each of Château Margaux 1953 and Chambolle-Musigny 1969. The price for four people: $125.

In the postwar years and well into the 1970s, Le Chambertin was one of many small French restaurants in the West Forties and Fifties. They were fashioned after working-class bistros in France and were run mostly by Normans and Bretons, many of whom had worked on French ships. With the French Line piers only a few blocks away, the West Side became a small but thriving French community.

Mr. Daniel came to New York in 1950 from Bermuda, where he had been part of a team of cooks, waiters, and waitresses that reopened the Castle Harbour Hotel there. Most of the Bermuda crowd migrated to New York.

Mr. Daniel had become involved in the restaurant business as a teenager, when he worked in restaurants in Paris. In 1942, during the German occupation of the city, he was taken prisoner and was sent to Germany to work in war plants. Years later, rocking with laughter, he would tell how he sneaked from the workers' barracks one Christmas Eve, stole a pig from the Germans' kitchen, and cooked it for his fellow prisoners.

Back in Paris after the war, he worked at the brasserie that the author Georges Simenon used as the model for the Brasserie Dauphine, a favorite of his fictional Inspector Maigret.

Here in New York, Mr. Daniel worked for several French restaurants, including Les Pyrenées, before opening Le Chambertin in the late 1950s. He also bought four acres in Phoenicia, New York, in an area known then, as now, as the French Catskills. There, on weekends and vacations, he was the consummate farmer and would fill his station wagon with fresh produce to serve his customers during the rest of the week.

Dinner at Le Chambertin was a treat, but dinner at his home in the Catskills was always something special. Often there was game he had shot, and always vegetables from the garden. As for the wine, he would emerge from his cellar grinning happily and carrying a magnum of Burgundy, perhaps a 1959 Gevrey-Chambertin or a 1971 Corton. "Not so bad for a country dinner, eh?" he would say. No, it wasn't, particularly after the white wines, Puligny-Montrachet or Meursault Perrières, that had started the meal.

Mr. Daniel closed Le Chambertin on December 31, 1975, worn out, he said, from trying to maintain a classic French restaurant in a city where fast food seemed to be taking over. But retirement was not for him. At age fifty-five, he became a wine salesman. Ignoring warnings that his old friends—restaurant owners and bartenders—might be less congenial when he came calling with his wine list, he built a successful second career, selling in Manhattan and the Hudson Valley.

The idea of a moderately priced French restaurant like Le Chambertin, with a staff of twenty, all of them French except the busboy, and a classic menu and a decent wine list, is only a fond memory now. Of course New York is hardly lacking in good restaurants. But how many of them or their owners will evoke similar memories twenty years from now?

DECEMBER, 1998

# CHAPTER 7

# WINE FROM PLACES OTHER THAN CALIFORNIA AND FRANCE

Wine is made in the oddest places, like India, China, Mexico, and Peru. It's even made from grapes grown on apartment house terraces in Paris. Once a lot of wine was made in England. For a long time there was none; then about fifty years ago, a movement to restore old vineyards and start some new ones got under way. Breaky Bottom is a product of that revival as a visit back in the 1980s shows.

Austria is almost fully recovered from a wine scandal back in the 1980s and is rivaling Germany and Alsace with its exquisite rieslings and grüner veltliners. Chile is booming and Argentina, that sad giant with infinite potential, is striving to catch up with some excellent wines of its own.

Spain is hardly a remote or unexpected wine source, but new wine areas have started up, including Priorat, south of Barcelona, and Toro, west of the exciting Ribera del Duero region. Portugal is making attractive table wines, and even the stolid Swiss are coming up with interesting wines, mostly white.

Most of us will rarely see these wines, much less taste them. But it's good to know that wherever you travel these days, you are probably not too far from someone who is crushing grapes and fermenting the juice.

# THERE IS INDEED AN ENGLISH WINE COUNTRY

I t's out here somewhere," the driver said, peering through the rain at the muddy dirt road. All we could see, except for the downs undulating off into the early spring mist, were fat Herefords gazing impassively at the sight of a taxi pretending to be a Land Rover.

"Ah, there it is!" the driver exclaimed, pointing down into a hollow so deep that the farm buildings in it were hidden by the slope dropping away from the side of the road. Cautiously, we began our descent. We had found Breaky Bottom.

This was, improbably, wine country—English wine country. I had come to Lewes by the ten A.M. train from Victoria Station. The suburban sprawl of London fell away quickly after Croyden, and in little more than an hour I was in the heart of Sussex. I was searching for Breaky Bottom, one of England's better-known small wineries, and Peter Hall, its creator and owner.

Mr. Hall, tall and rustically elegant in wellies and a scruffy silk neckerchief, met the cab carrying a newborn lamb. After a brief visit to the shed where another lamb, just born, was trying its shaky legs, we entered his home, a converted 1820s shepherd's cottage. Heat—a little—comes from the coal range in the kitchen, so we pulled up two chairs and sat facing each other in front of it.

Mr. Hall extracted a tin of Old Holborn tobacco and rolled a cigarette. "I was born in London," he said, "in Notting Hill. I always wanted to be a farmer, but I couldn't bloody well plow up Notting Hill." He studied agriculture at Newcastle University, and in the early 1960s took a job on a farm outside Lewes. "It was my postgraduate course," he said.

One day, the farmer sent him out to find some stray ewes and he stumbled on Breaky Bottom (*breaky* means "bracken" in the local argot). "It's not Wuthering Heights," he said, "but it's remote and lonely. I loved it from the first."

The farmer, who owned the cottage, let him stay in it. "I shoveled out tons of cow dung—they were the only creatures who used the place—and I mended the windows," he said. "And then I raised pigs for ten years. I am grateful to those pigs; they enabled me to raise my family and they paid for my first vines."

One day in 1974, he saw an ad for some vines. The English wine revival was in full cry and some of the best-known vineyards were only miles away. Mr. Hall liked the idea of growing grapes and bought the vines, Müller-Thurgau, a German crossbreed intended for northern climates. He also planted nine rows of seyval blanc, a French-American hybrid currently popular with East Coast growers in the United States.

After seven years, he pulled out much of the Müller-Thurgau, which produces a soft riesling-style wine, and added more seyval. It makes a fresh acidic wine close in style to Sancerre and other French white wines based on the sauvignon blanc grape. "Didn't want to go all Germanic," said Mr. Hall, whose mother was French. Today he makes both wines from about seven acres of grapes. Oz Clarke, wine critic for *The Daily Telegraph,* thinks Breaky Bottom wines are unquestionably England's best.

The Romans made wine in Britain. So did the first Benedictines, whose appearance in England may actually have overlapped the final days of the Roman epoch. Monasticism waned in the eighth century, and so presumably did winemaking.

Both flourished again after the Norman invasion. The new abbots were French and were used to good wine. The *Domesday Book,* compiled in 1084, listed thirty-eight English vineyards. The last item in a copy of the Anglo-Saxon chronicle tells of Martin of Peterborough, "a good monk and a good man" who "admitted many monks and planted vineyards."

Had not Eleanor of Aquitaine dumped Louis VII for the handsome Harry Plantagenet, Henry II, there is every reason to believe the English vineyards would have continued to expand. But Eleanor's dowry included Bordeaux. For the next millennium, the English would call claret their own.

Vines continued to be planted in England and Wales, and wine was made, but haphazardly and only for private use. Two diseases from America, powdery mildew and downy mildew, first detected in the nineteenth century, effectively killed off even the amateurs' vines.

The revival of English wines began tentatively in the 1930s and took off soon after World War II, when Edward Hyams planted a vineyard near Canterbury, in Kent. The first modern viticultural research station in Britain was opened at Oxted, Surrey, in 1946.

Since then, most English wines have been made by what the London wine writer Jane MacQuitty calls "eccentric gentlemen farmers and infatuated weekend winemakers." That is changing. There is a 65-acre vineyard in Wellow, in Hampshire, and the 250-acre Denbies estate in Surrey.

Two long, dry, and sunny summers in 1989 and 1990 gave rise to false hopes that the greenhouse effect had once again made England wine country, the way it had been under the Emperor Probus. In 1989 wineries in England, Wales, and Ireland produced 270,000 cases of wine, an all-time record. And one that is liable to stand. In 1991 harsh spring frosts, rain, a late summer, and a miserable autumn doomed all but a few vintners to thin, acidic, low-alcohol wines.

The country's entry into the Common Market makes its vinous future even more problematical. Overproduction of wine is a major problem in the European Community, which, pressured by France, looks askance at wines made from hybrid grapes like the seyval. Hybrids are banned in France. The Community will admit only what it terms "quality wine," and that eliminates any made from hybrid grapes. Britain's Agriculture Ministry is fighting those rules.

The Common Market notwithstanding, modern viticultural practices and carefully selected grape clones have brought winegrowing back to England. Some farmers have uprooted hops and fruit trees to make room for grapes. But only the most optimistic vintners expect English wine to ever be more than a marginal agricultural endeavor.

Peter Hall has no illusions about the harshness of his calling. Of the fifteen vintages he has worked, fully a third produced no wine at all. In 1989 he made twenty-seven thousand bottles; in 1990, twenty thousand, and in 1991, eighty-five hundred. "I can actually get by on ten thousand," he said. "Of course, I haven't had a holiday in twenty-five years and I work seven days a week."

Well then, one has to ask, why do this?

"Because," he replied simply, "I am stubborn and tenacious. And I happen to love what I am doing."

His sheep bring in some money, especially the lambs, which he sells for

meat. Picking up one of the bleating newborns on a walk around his property, he says to it: "You're going to be delightful with a touch of garlic and
some rosemary."

APRIL, 1992

## ON THE DANUBE,
## A QUIET REVOLUTION

Back in the early 1980s, I made a trip to Vienna to cover a conference
on arms control. I've forgotten the conference, but I do remember
*The Marriage of Figaro* one night at the Staatsoper. I remember, too, how the
aroma of chocolate filled the Sacher Hotel each morning, from all those
Sacher tortes baking downstairs.

And I remember my first great Austrian riesling. It was, as I recall, from
Willi Brundlmayer, a fine winemaker in the Wachau region, some forty
miles west of Vienna. Neither German nor Alsatian in style—it had their
richness and depth without their floweriness—it was a whole new experience.

Since then, there haven't been many opportunities to drink wines like
Mr. Brundlmayer's. For years, Austria was the forgotten nation when it
came to wine, almost nonexistent in the wine shops and restaurants of
America. And when anyone did think of Austrian wine, it was of the *Heurigen,* the young wines served in the little cafés in the woods around Vienna
where the tourists joined the locals to eat sausage and quaff pitchers of
frothy, forgettable white, most of it made from a uniquely Austrian grape,
the grüner veltliner.

But that's beginning to change, and it's because of a simple fact: In recent years, there has been a revolution in Austrian winemaking, and the
wines today are far better than they used to be. "For uninhibited enthusiasm and experimentation," Hugh Johnson wrote recently, "her producers
today rate somewhere between Australia and New Zealand."

The principal innovation, aside from technical advances in the vineyards
and cellars, has been a dedication to the riesling grape. My Brundlmayer
riesling in 1983 was, I realize now, something of a rarity. Today there are
dozens of fine rieslings coming from Austria, wines to match the best of

Alsace and Germany. And startlingly enough, there is a new effort to produce good red wines.

The rieslings and others are turning up in wineshops and in restaurants as diverse as Union Pacific, Judson Grill, and Picholine, which has several by the glass. Danube, David Bouley's Austrian restaurant in TriBeCa, has a list of 135 Austrian wines, divided into nineteen regions.

It isn't as if Austria were new to wine: Vineyards were planted along the Danube thousands of years ago. And the Austrians are wine drinkers. They are only twenty-first in the world in terms of the amount of land they have in vineyards, but they are eleventh in the world in per capita consumption (about eight and a half gallons a year, compared with less than two gallons in this country).

Just about all Austrian wine is produced in the eastern part of the country, with many of the best vineyards literally in the outskirts of Vienna. The best regions are the Burgenland, which starts a few miles southeast of the capital and extends along the Hungarian border, straddling the Neusiedler See, a shallow lake; the Wachau, along the Danube to the west, which has been called the Rheingau of Austria; the Kamptal and Kremstal areas, near the Wachau; and the Weinviertel region, to the north and east of Vienna, abutting the Czech and Slovakian borders.

The principal Austrian white wine grapes are the grüner veltliner, the welschriesling, the Müller-Thurgau, and of course the riesling, which accounts for only a fraction of total production.

The grüner veltliner, aside from its association with the Vienna *Heurigen,* turns out to be the source of some excellent white wines that, to my taste, have a resemblance to good, dry chenin blanc. Welschriesling, which is not really a riesling at all, and Müller-Thurgau are used mostly for cheaper commercial wines. Chardonnay, something of a newcomer, is growing in popularity, although the few I've sampled are burdened, like so many California chardonnays, with too much oak.

The principal red grapes are blauburgunder (the German name for pinot noir), blaufrankisch (thought to be from the gamay family; some of it is grown in Washington State, where its known as *lemberger*), and zweigelt; there are also small plantings of cabernet sauvignon and nebbiolo in some areas.

The Austrians, many of them, had always been as casual about making wine as they were about drinking it. Historically, thousands of gallons of cheap wine had been imported from Eastern Europe and fobbed off as domestic in Austria. Similarly, Austrian wines were often shipped off in bulk to be sold in Germany as domestic wine.

In 1985 Austria was rocked by a wine scandal. A number of major producers were found to have doctored thin white wine with diethylene glycol to give it body. So severe was the world's reaction that the Austrian government revised all its wine control laws, and producers began to take their work more seriously.

"It was a good thing, in the long run," Bertold Salomon of the Austrian Wine Marketing Service said in an interview from Vienna. "With the big producers and their cheap wines out of the picture, the small fine winemakers saw their opportunity. It was a chance to go for quality and they took it—very successfully, we think."

They did what all serious winemakers do, given the chance. They concentrated on noble grapes, particularly riesling. They cut yields to add intensity to their grapes, and many of them began fermenting and aging their wines in expensive new wood.

Austrian food and wine are enjoying a modest vogue in this country, led by Austrian chefs like Wolfgang Puck at Spago in California and Mario Lohninger, the *chef de cuisine* at Danube in New York. Austrian wine stewards like Alexander Adlgasser at Danube and Eduard Breg at the Tavern on the Green are dedicated to promoting the wines of their native land.

It's difficult to pinpoint the beginning of America's new interest in Austrian wine and food. A restaurant called Vienna 79 enjoyed considerable success in New York for a few years, before it closed in 1986. No serious Austrian restaurant followed it until Danube. Mr. Bouley, who once worked at Vienna 79, is leading the resurgence in New York, but his is just one of many restaurants offering respectable selections of Austrian wines: They include Jean-Georges, Windows on the World, Grand Central Oyster Bar, and Montrachet. On the West Coast, the Water Grill in Los Angeles and MC2 in San Francisco also feature Austrian wines.

To learn how the new Austrian wines worked with food, I asked Mr. Adlgasser to pair a half-dozen of them with Mr. Bouley's and Mr. Lohninger's food. For the whites, he chose a 1997 grüner veltliner, a Durnsteiner Kellerberg from the Wachau; a 1997 riesling *"alte reben"* ("old vines") from the Schloss Gobelsburg, in the Kamptal; and a 1995 chardonnay from a Viennese producer, Weininger.

The grüner veltliner started out with a lively, fresh, spicy taste. The riesling seemed closed in, and the chardonnay attractive but too oaky. Within half an hour, the riesling began to come into its own, and by the end of the meal, of all the wines, red and white, it was easily the best—neither as flowery as an Alsatian riesling nor as sweet as a German version, but a rich, subtle wine distinguished by its remarkable restraint.

Of the two reds, both 1996s and both from the Neusiedler See area, I preferred the blaufrankisch from the Weingut Romerhof over a blauburgunder Junger Berg from a producer called Umathum.

The finale was a dessert wine, a 1997 beerenauslese riesling from Aloïs Kracher, a producer from the Burgenland on the Neusiedler See with a reputation for making great late-harvest sweet wines. While no match for the best German beerenausleses, this was a wine of considerable power, with plenty of acid to balance its natural sweetness.

Conclusion: Austria is a treasure trove of fascinating undiscovered wines, some as good as any in the world, all of them intriguingly different.

One of Austria's more innovative winemakers is Willi Opitz at Illmitz, on the Neusiedler See. Rather than just advertising his wines, he has produced a CD called *The Sound of Wine,* a recording of noises his wines make while fermenting. And with no apologies to Robert Mondavi, he calls one of his favorite wines, yes, Opitz 1.

NOVEMBER, 1999

## ARGENTINA IS COMING OUT OF ITS SHELL

The Argentines are a proud people, not given to hiding their many assets. Except when it comes to wine.

Argentina is one of the great wine-producing nations, right up there after France, Spain, Italy, and Russia. But hardly anyone knows this. Well, maybe a few people do, but they are mostly in Argentina. Why? Because it was a closed-in country, an economic and political basket case, for forty years. Isolated by domestic politics and killing tariffs, the Argentines had little choice but to go their own way. Until now. Thanks to a stretch of stability that looks as if it just might last, they are catching up fast, and nowhere faster than in the wine business.

The wine trade in Argentina has a long history. The first commercial wineries were developed by missionaries in the sixteenth century, using vines brought over the mountains from Chile and Peru. Wine was an artisanal trade until the arrival of hundreds of thousands of Italian immigrants in the late nineteenth century, most of them amateur winemakers, more than a few of them professionals. Many settled in Mendoza, six hundred miles inland from the Atlantic and to this day a city more Italian in feeling than parts of Italy itself. Those immigrants and their children made Mendoza, both the city and the province, the center of the Argentine wine industry, producing 70 percent of the country's wine.

Substantial vineyard areas can also be found five hundred miles to the south at Río Negro, in Patagonia, and some seven hundred miles to the north in Salta and in Jujuy Province, not far from the Bolivian border.

While Argentina's wine production has dropped in recent years, it remains immense, but almost all of it is consumed domestically. Even so, yearly consumption has dropped precipitously, from about twenty-five liters a person in the 1960s to ten liters today. Like the rest of the world, Argentines have discovered beer and soda, and so the industry has been forced to develop an export market.

Though Argentine vineyards still supply millions of gallons of cheap wine to northern Europe, the home market in the cheap stuff has all but died away. The world is drinking less but drinking better wine, and that's what the Argentines are beginning to make.

As in Chile, there has been a trend toward expensive, limited-production wines. While I'm not sure who will pay $50 for Catena's Algelica Vineyard Malbec or Trapiche's Iscay malbec-merlot blend, there are dozens of new, reasonably priced releases that are worth sampling.

Under the governments of Juan Perón and a succession of military juntas, the wine industry operated in virtual isolation, not unlike the South African wine industry during apartheid. Not until trade and political barriers were lowered did winemakers or consumers have an idea of how winemaking had improved worldwide. The growers began a program of pulling out old vines and replacing them with fewer but better European varieties like cabernet sauvignon, merlot, and chardonnay.

Malbec, once an important Bordeaux grape and still the favorite grape of Cahors, in southwestern France, is also Argentina's most prominent red wine grape. At its best, it makes a better wine in Argentina than it does in France.

Mostly because of political and economic difficulties in the 1970s and '80s, Argentina got a slow start in the race with Chile for a share of the expanding world market for fine wine. But there is a widely held perception that Chile has yet to fulfill its early promise as a source of truly great wines and that Argentine winemakers have been steadily narrowing the quality gap between the two.

What the Argentine domestic market consumed in vast quantities was mediocre wine made from the pedestrian criolla grape. Thousands of acres of criolla were ripped out. It had been replacing malbec, which is less economical to grow. But the malbec produced the country's best wine, so with the possibility of a new export market looming, hundreds of acres of malbec were replanted.

The return of a degree of economic stability has lured investments. Pernod-Ricard from France, Allied Domecq from Britain, and Kendall-Jackson from California have been putting money into Argentina that, a decade ago, might have gone into Chile. In fact, even the Chilean winemakers are getting into the game: both Concha y Toro and Santa Carolina have invested in Argentina.

Until recently, rosé wines were more popular than reds and whites in Argentina. But the recent interest in better wines, particularly among the new middle class, has shown a preference for whites. It is a preference unlikely to spill over into the export market because Argentine whites are not a distinguished group. The torrontés grape, a Spanish import, is Argentina's unique white wine, the Spanish having pretty much given up on it. It makes a fruity wine with nice structure, but the Argentines, like everyone else, seem drawn to chardonnay. And they seem to be making the same boring chardonnays as almost everyone else.

Argentina's future would seem to be tied closely to red wines. Until recently, even its best were not contenders. They were heavy and unbalanced. Which, not coincidentally, was the way Argentine consumers liked them. Fortunately, Argentine winemakers are eager to learn international styles.

Along with the venture capital has come a small army of French, Italian, and American winemakers. A British wine writer, Fiona Beckett, has said that Bodegas Norton in Mendoza is working with six clones of cabernet while another producer, Leoncio Arizu, is growing eight varieties in an experimental vineyard, with each variety planted six different ways. The results have been impressive—and not just from those forward-looking producers.

Bodegas Etchart, now owned by Pernod-Ricard, has two lines of inexpensive wines that are remarkable values. From Mendoza vineyards, the Río de Plata line offers four wines: the usual suspects, cabernet sauvignon, merlot, and chardonnay, plus malbec, each at about $7. The wines are one-dimensional but full-bodied and fruity, with the malbec and the cabernet out front.

From Andean vineyards at Cafayate in the north, the Etchart wines include a delicious cabernet sauvignon at about $9 and, for anyone who wants to try something different, a torrontés white at the same price. These wines are all 1999s.

Navarro Correas, one of the oldest and most distinguished Mendoza wineries, has been in the American market for many years. Its Malbec and its Private Selection Cabernet Sauvignon, at about $12 each, are good buys. I tried the 1994s; younger wines should be available soon.

APRIL, 2000

## SPAIN'S NEW VINTNERS LEAVE RIOJA IN THE DUST

As long as anyone can remember, red wine in Spain has been synonymous with Rioja, the fertile valley of the Ebro River in the northernmost reaches of the country, not far from the western Pyrenees and the French border. In recent years, however, Rioja has begun to confront some formidable competition.

The Penedès region, southwest of Barcelona, was Rioja's biggest challenger until recently, thanks mostly to one producer, Miguel Torres. But now other regions of Spain are clamoring for attention and, in some cases, making wines that prove they deserve it.

From the Old Castile region in the north-central part of the country, the quality of wines from the Ribera del Duero and Sardon del Duero vineyards has astonished connoisseurs. Once known principally as the source of Vega Sicilia, Spain's most famous and most expensive red wine, the Ribera del Duero is considered by many experts to have surpassed Rioja as the country's most important red-wine-producing area.

In Navarre, producers like Chivite have revived old winemaking skills to produce wines that rival and in some instances surpass those of neighboring Rioja. Good wines now come from Valencia, from Valladolid, and even from La Mancha, once a jug wine wasteland, where Carlos Falco uses techniques he learned at the University of California to make fine cabernet sauvignon.

And then, in the rugged mountains west of Tarragona in southern Catalonia, there is Priorato, the most exciting new source of fine wines on the Iberian peninsula. In the last decade, a steady stream of enological pioneers and iconoclasts have gone there to transform the inhospitable countryside into a wine region to rival the best in Europe.

Spanish wine has had two great moments in the sun. The first was in the 1870s, when phylloxera devastated the vineyards of France. Bordeaux vintners flocked to Rioja, first to buy wine and then to make it there. Eventually Spanish vineyards, too, were laid low by phylloxera.

The second surge of interest came with the end of Spain's political and cultural isolation following the death of Francisco Franco in 1975. Millions of visitors rediscovered the delights of Spanish food and wine. Most of the new developments in Spanish winemaking are a continuation of the modernization of the industry begun in the late 1970s and early '80s.

Some Rioja producers, of course, continue to produce wines that are true expressions of the tempranillo grape. In the past, too many of these wines were burdened by the cloying effects of too much American oak. The Contino winery in Laserna knows how to handle oak and produces some luscious, big red wines. Compania Vinicola del Norte de Espana (Cune), in Haro, offers plenty of oak, but some of its best wines, like the Imperial Reserva, handle it nicely.

American wine drinkers have taken enthusiastically to the Conde de Valdemar wines of Bodegas Martinez Bujanda. These wines are light, refreshing, and fruity and have virtually nothing in common with the classic Rioja style. My friend Gerry Dawes, a wine writer who is probably America's premier expert on Spanish wines, hates this stuff, but he is a traditionalist. I think it's tasty. Two other Rioja names to watch for: Remelluri and Bodegas Muga.

In Navarre, Bodegas Julian Chivite is the best-known producer, and its best-known wine is, of all things, a rosé. That wine is called Gran Fuego, and it's probably a good introduction to this producer's wines. I would prefer the Collection 125 Gran Reserva, a tempranillo wine with some cabernet and merlot, but any Chivite wine is worth sampling.

Torres, in Villafranca del Penedes, makes a huge variety of wines from traditional Spanish grapes and some not-so-traditional ones, including gewürztraminer and sauvignon blanc. Recently I tasted several vintages of Torres's top-of-the-line Mas la Plana Black Label Cabernet and was very impressed. This is almost always a big, powerful wine, good for drinking five years after the vintage but with tremendous aging potential. It is about $30.

At the lower end of the price scale, the simple Sangre de Toro, made from grenache and carignane, is, at $7, invariably one of the best buys in any wineshop.

In the Ribera del Duero, the story of Alejandro Fernandez's ascension from winery-equipment salesman to one of Spain's best-known winemakers is striking, but no more so than some of the wines he makes. Pesquera, his principal wine, is deep and dark, yet usually smooth and elegant. Mr. Dawes has aptly compared a Pesquera Gran Reserva to a great Pomerol.

At their best, the wines coming down the mountain trails from Priorato are blockbusters. They bring to mind some great Hermitages, the Barolos of Bruno Ceretto, Guigal's Côte Rôties, and a couple of Helen Turley's zinfandels. Garnacha and carinena (grenache and carignane) are the principal grapes of Priorato, with an occasional touch of cabernet for good measure.

Everything is small in scale in Priorato except the prices, which can reach $200 a bottle, and the alcoholic content of the wines, which often tops 15 percent. The vineyards are small, the cellars are small, and so is the quantity of wine they make.

The Clos de l'Obac from Costers del Siurana, a blend of cabernet, merlot, and traditional Spanish grapes, is a lighter-style Priorato and a good introduction to these wines. Alvaro Palacios's Clos Dolfi is more in the classic vein: dark, intense, and muscular. A four-year-old Clos Dolfi probably needs five more years of bottle age. Mr. Palacios's L'Ermita sells for about $200, and his Clos Dolfi for about $50. Both are rare. But then so are the less expensive Prioratos like Mas Igneus Barranc dels Closos. It should sell for about $15, if you can find it.

MAY, 1999

## FINDING PORTUGAL

I love white Portuguese wine more than claret," Jonathan Swift wrote to his beloved Stella. "I have a sad vulgar appetite."

Not a man to mince words, Dean Swift. But one suspects he was also a bit of a snob. In his day, in the first half of the eighteenth century, England drank five times more Portuguese wine than French. It was the *vin ordinaire* of that epoch. Thanks to Hanoverian import duties and the Whigs who imposed them, claret—red Bordeaux—was more expensive than caviar is today; only aristocrats could afford it.

Once relations with France had improved, the English abandoned Portuguese wines. Except, of course, for Port, which they concocted as an early substitute for central heating. Portuguese wines have never quite recovered from the snub. They have been dismissed as second-raters for two centuries. And in truth, they've often lived up—or, rather, down—to their reputation.

There is an interesting paradox associated with Portuguese wine. Ask a fellow who drinks an occasional glass of wine if he likes Portuguese wine—other than Port, that is—and you will get a blank stare. Ask him if he likes Lancers or Mateus Rosé, and his face will light up. Americans who care little about wine drink millions of gallons of these pink, fizzy commercial blends. Meanwhile, the best Portuguese wines go begging. Wine enthusiasts, who take pride in their eclectic tastes, ignore them.

The modern equivalent of Jonathan Swift's white wine is *vinho verde,* or green wine. It's not really green, of course, but has a slightly greenish tinge. At its best, it's light, fresh, sharp, and slightly fizzy. At its worst, it is sweet, sulphured, and gets its fizz from carbon dioxide. There are some single-estate vinho verdes, but most of the wine is made by cooperatives in the Minho region in northern Portugal, between the Douro River and the border with Galicia in Spain. The cooperatives buy both grapes and wine from small producers and blend it. Vinho verde is a wine to be consumed young, preferably close to where it's made. It matches well with the good fish caught along the Portuguese coast.

There is also a red vinho verde that is drunk mostly in Portugal and usually exported only to countries with large overseas Portuguese colonies.

The grapes included vinhao, azal tinto, and espadeiro. How do the Portuguese justify the name? They don't bother to try.

The largest wine region is the Estremadura, but the best reds come from the Dão, from Bairrada and from Colares. The Dão, which is south of the Port-producing Douro region, was Portugal's first government-authenticated red wine district, the first to produce serious *vinho maduro,* the term used for matured wine. Most of the Dão wines are merchants' blends bearing brand names rather than those of the grapes or individual producers. The wines are usually sold when they are about four years old. Traditionally, they are hard wines that need years to mature. Modern vinification methods have made them more drinker-friendly.

Wine has been made in Portugal at least since Roman times, but the Bairrada was not declared a demarcated region until 1979. It lies between the Dão and the Atlantic Ocean, and the sea mists, by moderating the normally torrid Portuguese climate, contribute much-needed acidity to the wine. Portuguese viticulturists say the basis of the Bairrada wines' quality is its principal grape, the baga, which they say gives the wine color, fruit, and longevity. Caves Alianca is a Bairrada producer whose label is often seen in this country.

Some of Portugal's more interesting red wines come from Colares, west of Lisbon on the coast. One could say literally on the coast, because the vines grow on the beach, in the sand. The vines creep along the ground behind low stone walls and woven straw fences that protect them from the wind. The wines are intense, almost black in color, and take years to mature. When they do, they are said to be Portugal's best red wines.

Lisbon is virtually surrounded by vineyards. Directly to the west is the Carcavelos area that produces a sweet amber-colored wine. Bucelas, north of Lisbon, produces a pleasant but inconsequential white wine. Southwest, across the Tagus bridge, is the Setúbal area, which produces one of the finest muscat wines in the world. Called, not surprisingly, Moscatel de Setúbal, it is a wine that lasts for many years, improving in taste and complexity as it ages.

While most Portuguese table wines are commercial blends, there are a few outstanding single estates. One of them, Bussaco, comes from the Bairrada region and is made only for the Palace Hotel in Bussaco, which is south of Oporto, near the spa resort of Luso. The wine, red and white, is made by the hotel's manager, José Rodrigues dos Santos, and is aged and bottled in the hotel's cellars.

Barca Velha, "the old barge," is a red wine made from grapes grown in the remote Alto Douro, where the grapes for Port are also grown. The estate is owned by descendants of Dona Antonia Ferreira, the doyenne of the Ferreira Port concern, and is vinified by Ferreira winemakers. Each vintage is aged up to two years in new Portuguese oak casks. It can take years to mature in the bottle, and it is released only when the man who created the wine in the 1950s, Fernando Nicolau de Almeida, decides it's ready. Only ten vintages have so far been released: 1953, 1954, 1957, 1964, 1965, 1966, 1978, 1981, 1982, and 1983. Lesser vintages sometimes turn up in a wine called Reserva Especial Ferreirinha.

In Portugal, as in almost every other important winemaking region in the world, the famous Bordeaux grapes like cabernet sauvignon, merlot, and cabernet franc are being planted alongside the indigenous varietals and are giving them stiff competition. Bacalhoa, a red wine from a vineyard in Setúbal that is owned by Americans, is made entirely from cabernet sauvignon and merlot and has been rated one of Portugal's best single-estate wines.

Whatever it does for local pride, using the Bordeaux varieties is a boon for wine writers—wine writers, that is, who know nothing of the local varieties like baga, tinta pinheira, bastardo, trincadeira, tinta roriz, touriga francêsa, touriga nacional, alicante, and periquita. To name a few. So as they say in Portugal, *Saude.*

SEPTEMBER, 1991

## THE SWISS PARADOX

Switzerland is wine country, too. Wine is made in nineteen of the twenty-three cantons. It is also imported in vast quantities, mostly from France and Italy. And it is drunk everywhere, from Geneva in the west to Rorschach in the east and from Zurich in the north to Zermatt in the south.

The Swiss drink about 80 million gallons of wine a year, which is not bad for a small country. We drink about five times as much, but there are forty times more of us than there are of them.

But their vineyards produce, on average, only about a third of what they consume. One would assume that the Swiss regularly use up all their wine. Not so. There are times—and this is one of them—when the Swiss can't give their wine away. It simply costs too much to make. Swiss winemakers are the highest paid in the world, and their wine prices show it.

Switzerland has reduced its vineyards by almost 75 percent over the last century. The irony is that efficient viticultural practices enable the growers to make almost as much wine on the vastly reduced acreage.

For a number of years, the Swiss government sponsored a restaurant in New York City that featured Swiss wines. The retail trade never showed any interest. A few Swiss-style restaurants feature bottles from the old country, but the idea never seems to go any further. They are, for the most part, too expensive for what they are.

And what are they? Generally, good white wines made in styles very different from those we get from France, Italy, or Germany. Very different, yes, but not necessarily better. Current worldwide overproduction has made commercial white wine from France and particularly Italy relatively inexpensive. The Swiss cannot compete. With their reds, it's more a question of quality. There are fairly good Swiss red wines, but it is not just happenstance that the Swiss are the largest importers of Beaujolais in the world and among the largest importers of Burgundy.

Where wine is concerned, Switzerland is a paradox: a wine-producing, wine-drinking country that refuses to consume its own product. Tough laws limit the growing of grapes. If vineyards outside of the stipulated viticultural areas are found to exceed four hundred square meters, which is just about enough for home consumption, they may be destroyed by helicopters. Unless there has been a very bad crop in Switzerland, no bulk white wine can be imported and the amount of bulk red is severely limited. Various efforts have been, and continue to be, made to control the amount of wine released to the market, and to get importers of French and Italian wine to absorb overproduction of Swiss wine.

Geneva is in Switzerland, but only just. As anyone who has flown in and out of the Geneva airport knows, if you leave by the appropriate door you are in France. The Gallic influence is quickly evident in the wine lists of the better restaurants. Finding a good Swiss bottle takes a bit of effort. Not so in Lausanne, an hour's drive to the east along the north shore of Lake Geneva. Most of the attractive lakeside restaurants proudly feature the local wines. As well they might—fully 40 percent of all Swiss wines come from vineyards planted east and west of the city along the northern shore of the lake.

This winegrowing region which stretches from a short distance north of Geneva to Montreux at the eastern end of the lake, produces some of Switzerland's best wines. One of the most famous vineyards is at Dézaley, which many Swiss say is their best. Others bear such obscure names as Saint-Saphorin, Rivaz, Riex, Lutry, and Cully. Actually, the vines continue down past Montreux and follow the northern shore of the Rhône as far as Visp, the jumping-off spot for Zermatt. This is the Valais, the oldest vineyard area of Switzerland and probably one of the oldest in Europe. There is an annual tradition here of lugging eight-gallon barrels of the new wine up into the Alps that dominate the valley. There, above the clouds, it is stored in cellars for ten or fifteen years. What results is a hard white wine called Vin de Glacier.

One of the most attractive Swiss wine regions runs along the northern shore of Lake Neuchâtel and Lake Bienne, about halfway between Geneva and Basel. The Neuchâtel wines are exported more than the wines of Bienne, but the neatly tended vineyards along the shore of Lake Bienne are more inviting. The vineyards slope up from the road that parallels the lake and the railroad that parallels the road. On the water side of the busy roadway, delightful-looking inns and restaurants jut out over the water, their signs proclaiming almost constant tastings of the local wines.

Bienne marks the dividing line between French- and German-speaking Switzerland, but its wines are in the French style. The chasselas and the pinot noir are the predominant grapes grown here.

Wines are made in the German-speaking side of the unofficial border, too. Pinot noir is the most popular grape, but there are some white grapes that are vinified in the German style.

In the Ticino, Italian-speaking Switzerland, all the talk these days is about merlot, which was introduced here from France, where it produces the great wines of Pomerol, among others. It has edged out the nostrano, the local red-wine grape, which produces a wine that only a Ticinese could love.

It's worth noting that the chasselas, a grape held in relatively low repute elsewhere, is much loved in Switzerland. It produces quality white wines with aromas and flavors that change from one vineyard region in the country to another. This is something it seems incapable of doing elsewhere.

FEBRUARY, 1984

## ANGELO GAJA

You're inching down the road from Asti to Alba, looking at the vineyards and wondering idly if the fellow creeping along in front of you actually knows about second and third gear. But the sun is shining; the birds are singing, and the grapes look good. So you relax.

Then a dark spot appears in the rearview mirror. It grows swiftly until a black shape hurtles by you and disappears around the first turn ahead. It is Angelo Gaja, Italy's best-known winemaker, in his new BMW, off to visit one of his vineyards. "I guess I am always in a hurry," he says later.

And so he is. On the road five, even six, months a year selling his wines, he puts in twelve-hour days when he's home to make up for the time he spends in airplanes. Not only does he make some of Italy's—some say the world's—finest wines, he also imports first-growth Bordeaux, famous Sauternes, and a line of top California wines. His company is the exclusive importer in Italy for the wines of the Domaine de la Romanée-Conti, in Burgundy. And as if that were not enough, he imports and sells fine china and crystal.

But it is by his own wines that the world knows Angelo Gaja. The son and grandson of Piedmontese winemakers, in less than two decades he has forged such a reputation for excellence that connoisseurs unhesitatingly rank him with the world's best. Which, by the way, is pretty much where he ranks himself. It's not just by chance that he sells only the best French and California wines; it's called positioning.

The Gajas came from Spain in the seventeenth century. The present business was founded by Angelo's great-grandfather Giovanni, a grape grower who decided to start his own winery. Angelo, who studied winemaking at the Enological Institute in Alba and at the University of Montpellier in France, and who has a degree in economics from the University of Turin in Italy, joined the business in 1961.

The Gaja catalogue includes Barbaresco, Dolcetto d'Alba, nebbiolo, Barbera, chardonnay, cabernet sauvignon, and sauvignon blanc. All Gaja wines are made from grapes grown in Gaja vineyards, of which there are fourteen spread over the chalky hills of the Piedmont in northwestern Italy. A vineyard purchased in 1988 produces Barolo, a wine Mr. Gaja's father made from grapes purchased from the same vineyard thirty years ago.

The keystone of the Gaja legend is his Barbaresco. There are four, the regular wine and three single-vineyard wines: Sori Tilden, Sori San Lorenzo, and Costa Russi. Another Gaja vineyard, Vignaveja, outside the Barbaresco commune, provides the grapes for his Nebbiolo Vignaveja.

Both his fellow Barbaresco producers and his customers were surprised when Mr. Gaja introduced his cabernet sauvignon and chardonnay, both outstanding wines, in 1983 and 1985, respectively. They have received considerable acclaim, the chardonnay in particular, but he insists that they are essentially stalking horses for his traditional wines.

"There are markets all over the world for chardonnay and cabernet," he said, "but many people are still hesitant to try our traditional wines. If I can show them what quality wines we make here, perhaps they will drink our Barbaresco, too.

"I am proud of those wines," he said of the cabernet and chardonnay, "but I have perhaps just a little larger space in my heart for the nebbiolo." Both Barbaresco and Barolo are made from the nebbiolo grape. The name comes from *nebbia*—"fog"—the early-morning fog that so often clings to the Alban hills.

The cabernet, called Darmagi, is 100 percent cabernet and was first planted in 1978. Mr. Gaja said *darmagi,* which means, roughly, "too bad," is what his father used to say when he looked at the alien cabernet vines in good nebbiolo soil.

The chardonnay is called Gaia & Rey. Gaia is the name of Mr. Gaja's oldest daughter; Rey is his grandmother's maiden name. A sauvignon blanc first released in 1988 comes from a tiny three-acre vineyard called Alteni di Brassica.

Mr. Gaja has his detractors; they call his wines idiosyncratic and above all overpriced, but there are not too many enthusiasts who will argue with *The Wine Spectator,* a prominent consumer publication, which once called his Barbarescos "the finest wines ever made in Italy."

Mr. Gaja attributes the quality of his wine to attention to detail: in the cellar, where he was one of the first to introduce aging in small oak barrels, and in the bottling room, where he uses specially designed bottles and what must be the longest corks ever squeezed into glass. But there is more. Not content to buy barrels from skilled *tonneliers,* he buys the staves, then dries them for three years outside his winery before having them assembled by his own barrel makers. The new barrels are steamed for an hour before they are put to work in removing 50 percent of his Barbarescos' natural tannin. One complaint about Barbaresco has always been the harshness of its tannins. The suppleness and elegance of Gaja wines come at least in part from these treated barrels.

"My goal is to control every step of my production," Mr. Gaja said, "vineyards, grapes, barrels, everything. In the future, I think more and more producers will insist on controlling all the details of their winemaking."

Even with all the controls in his hands, Mr. Gaja isn't always satisfied. He refused to sell his 1984 Barbaresco, some twelve thousand cases including the single-vineyard wines, under the Gaja label. He and Guido Rivella, his enologist since 1970, agreed it should be sold off in bulk. "It was not a bad wine," Mr. Gaja, "but it was not good enough for the prestige of Gaja. You see, another detail."

Does all this add to the cost? Well, yes, one could say so. The single-vineyard Barbarescos sell for about $90 a bottle and more. The cabernet sells for around $70, or a bit less than a first-growth Bordeaux like Château Margaux or Château Latour. The chardonnay is around $40. The nebbiolo and Dolcetto are usually under $30. Most mortals gasp at such numbers, but there is a constant shortage of Gaja wines. Total production is only twenty-two thousand cases a year, and his American importers believe they could sell it all themselves.

"Once high prices were reserved for the French," Mr. Gaja said with considerable satisfaction, "but no more. Small producers in other countries are now capable of competing with the best in the world."

AUGUST, 1989


## A GAMBLER'S LEGACY

The story of South African wine began on February 2, 1659, when Jan van Riebeeck of the Dutch East India Company wrote in his diary: "Today, praise the Lord, wine was pressed for the first time from Cape grapes."

Within one hundred years, South African wines were among the most prized in the world. Thomas Jefferson served them at Monticello, and when Louis XVI died, his cellar was found to hold far more of the legendary Constantia, from South Africa, than it did fine Bordeaux.

Now, more than three centuries since that first vintage at the Cape, South Africa is the world's eighth-largest wine producer, with double the production of Australia and New Zealand combined. And, with the end of apartheid, trade embargoes have been lifted, and South African wines are once again available in this country.

The major figures in the early history of South African wine were a group of French Huguenots who arrived in 1688. They settled in Franschhoek— "French Corner" in Afrikaans—and turned it into the region where some of South Africa's best wineries are found. But the real story of South African wine began early in this century with an American adventurer named William Charles Winshaw.

Bill Winshaw was born in 1871 in Kentucky. He ran away from home at age eleven, and while still in his teens, prospected for gold in the West. When that—literally—didn't pan out, he supported himself by gambling.

Winshaw eventually settled in New Orleans, where he paid his way through medical school, also by gambling. He soon returned to New Mexico to begin his medical practice, and there he met a Lieutenant McGuiness, who had come west to buy mules for the British army. Soon after, in 1899, Winshaw was aboard the *Larinaga,* using his physician's skills to care for some four thousand mules that Lieutenant McGuiness was shipping to the British forces in Cape Town.

Once rid of his four-legged charges, Winshaw joined the British forces and fought in the Boer War. He remained in South Africa after the war and set up a medical practice in Cape Town. After his practice was settled, Winshaw rented a farm near Stellenbosch and began to make wine—on the kitchen stove. His true career had begun.

Winshaw was soon making wine commercially. He imported Concord grapes from the United States, and in 1909, with about $2,500, he opened the Stellenbosch Grape Juice Works, where he produced grape juice and wine. Stellenbosch, nestled in the mountains thirty-five miles northeast of Cape Town, was then and still is the center of the South African wine industry.

The business grew over the next decade, and by the end of World War I,

Winshaw was a major figure in the South African wine industry. But two years later, he was bankrupt. The market for South African wine collapsed in the postwar depression and the Winshaw interests were declared insolvent in January 1921.

Broke and discouraged, Winshaw returned to the United States, but only briefly. By 1924, at the age of fifty-three, he was back in the East Cape making wine in partnership with Gideon Krige, a wine-and-spirits merchant and the owner of a Stellenbosch distillery. Winshaw got his financial backing from a longtime friend and bought out Krige for about $7,500. With his sons, Bill and Jack, he established the Stellenbosch Farmers' Winery. Through expansion and acquisitions, it would become the top wine wholesaler in South Africa, marketing about two-thirds of all wine consumed in the country.

Winshaw emphasized dry table wines, a style still dominant in South Africa. His greatest success, however, was a semisweet wine called Lieberstein, which was introduced in 1959. By 1964, domestic and foreign sales had reached 3 million cases. SFW, as the winery is known, still makes a variety of sweet and semisweet wines, but like most South African wineries, its reputation today is based on its dry reds.

The Winshaws took their company public in 1935. In 1960 they sold their controlling interest to South African Breweries but continued to run the company. Bill Winshaw remained at the helm of SFW until he retired in 1962. He died in 1967, at the age of ninety-six. In 1966 SFW acquired the famous Nederburg Estate in Paarl, a winery that celebrated its 200th anniversary in 1992. Paul Pontallier, the winemaker at Château Margaux in France, is a consultant at Nederburg, where incidentally, South Africa's biggest wine auction is held every year.

SEPTEMBER, 1992

## ON THE NORTH FORK, DREAMS OF NAPA

It's high summer out here on the North Fork of Long Island. The corn is sweet, the first peaches are in, and in the vineyards, the still-green merlot and cabernet franc are looking good. It will be long weeks, maybe two

months, before the pickers arrive and the air grows heady with the seductive aroma of crushed grapes.

But don't be deceived by the bucolic calm. The Long Island wine business is booming; it's almost giddy from its own success. Handsome new
wineries are opening, old ones are changing hands for astonishing prices,
new vineyards are springing up from Aquebogue to Greenport, and suddenly there isn't enough wine to go around.

What's more, these developments are not due to the hesitant little band
of artisans and amateurs who started it all in converted barns and potato
fields some thirty years ago, seeking a simpler agrarian lifestyle, so they
thought. The current scene is made up of self-assured professionals backed
by deep-pocketed investors who are also seeking a different lifestyle—as
owners of Napa Valley–like showcase wineries making prize-winning and
sought-after wines.

World-class wines may be a few years away, but the money, the talent,
and the will to make them are all in place. It is, everyone out here will tell
you, just a matter of time.

I'm inclined to agree, with a few reservations. Seven of every ten bottles
of wine Americans consume is California wine. California wines, for the
most part, are jammy, full-bodied, velvety, and relatively easy to drink. Fat,
buttery chardonnay is the paradigm. Long Island wines, at their best, are just
as intense, but are leaner, less voluptuous. Perhaps more elegant. They are a
different taste experience, and for many wine fans they take some getting
used to.

The North Fork is a narrow strip of farmland on eastern Long Island
between Long Island Sound and Peconic Bay, about seventy miles from
Manhattan. In 1970 it was on the brink of the same drab suburbanization
that had already overwhelmed Nassau County and much of western Suf-
folk. Then the affluent and photogenic Alex and Louisa Hargrave, children
of the 1960s, arrived, wanting to go "back to the land." They founded Hargrave Vineyard in Cutchogue, and almost miraculously, other vineyards
soon followed, bringing new life to the faltering agricultural economy.

Today vineyard land totals about twenty-eight hundred acres, of which
about twenty-two hundred are in production. Annual wine production, still
negligible, is estimated to be three hundred thousand cases, about the same
as for one medium-sized California winery. It's hardly enough to supply a
growing market and the hordes of tourists who descend on the tasting

rooms each weekend, particularly in the fall, when some of the wineries receive four hundred to five hundred visitors a day. Not long ago, the Long
Island wineries sold most of their production in their tasting rooms. Now,
with their business growing in many markets, tasting room depletions are
forcing some to charge for what they once gave away free. (Several of the
region's twenty-one wineries are actually on Long Island's South Fork, which
was once agricultural but is now given over principally to the Hamptons.)

If any North Fork endeavor symbolizes how the Long Island wine industry has progressed in less than three decades, it is the striking Italianate
Raphael winery going up on the Main Road in Peconic, a hamlet ten miles
east of Riverhead. It is a $6 million testimony to the faith that the owner,
John Petrocelli, has in the Long Island wine industry. A year before its
opening, it has already raised the bar on what is expected from Long Island's
vineyards and fledgling wine producers in the years ahead. No expense has
been spared by Mr. Petrocelli, a commercial builder in New York City.

At the same time, Raphael is based on what came before: the winemaker is Richard Olsen-Harbich, a veteran of twenty years in Long Island
cellars.

Raphael, named for the owner's father, an immigrant and home winemaker, will be a one-wine winery, and that wine will be merlot. Of the
sixty acres of vines planted, forty-five are merlot. Small amounts of cabernet sauvignon and other Bordeaux varietals are cultivated, but only to enhance the merlot. There is—hold on to your seat—no chardonnay. In fact,
there is no white wine at all, except a smidgen of sauvignon blanc for
friends.

It was in deciding on merlot, and only merlot, that Mr. Petrocelli and
Mr. Olsen-Harbich drew a figurative line in the sand to mark the spot
where the old Long Island wine industry ends and a new one begins.

Mr. Olsen-Harbich has a foot in both camps. A twenty-year veteran of
the Long Island wine industry who was born in Queens and studied viticulture at Cornell University, he started as founding winemaker at Bridge-
hampton Vineyards, worked with the Hargraves for a time, and spent
several years as a consultant to newcomers eager to start up wineries. But
the arrival of Mr. Petrocelli ratcheted up the stakes.

"Hit-or-miss experimentation, putting vineyards in and pulling them
out—those days are all behind us," Mr. Olsen-Harbich said. "We know this
is merlot country now."

Does this sound rather like Bordeaux? It does, and with good reason. Mr. Olsen-Harbich shares his Raphael responsibilities with Paul Pontallier, the managing director of Château Margaux, one of Bordeaux's legendary *premier-cru* wine properties. He has been a consultant to Raphael since 1996, when the project began.

"Our soil, our climate, both are Bordeaux-like," Mr. Olsen-Harbich said, "and our wines taste more like Bordeaux wines than California wines. We needed the Bordeaux influence. Before we planted a grape, Paul determined the vineyard layout and the spacing of the vines. He is here four, occasionally six, times a year, and we talk, sometimes daily, on the phone."

Mr. Pontallier is circumspect: "The main problem on the North Fork is to get the grapes ripe," he said by telephone from Bordeaux. "The region has an interesting combination of warm and cold temperatures, and the grapes must struggle to get ripe. Often they are not picked until November. Our merlot here in Bordeaux is three to four weeks ahead of the merlot on the North Fork.

"We are not going to make great wine overnight," he added, "but we will make very good wine. Of that I am sure."

Raphael, using Pellegrini's cellar, produced 200 cases in 1997 and about 700 of the 1999 vintage. The goal is around 10,000 cases. Compare that with Pindar Vineyards, the region's largest, which reports it produces more than 80,000 cases and plans to reach 150,000.

Even if its production is small, Raphael is a winery on the grand scale. It presents the challenge for new investors to top. Robert Entenmann, of the baked-goods family, who owns Martha Clara Vineyards in Mattituck, might be one of the first to do that. He has plans for a $4 million tasting room and a $2 to $3 million winery. Leslie Alexander, the owner of the Houston Rockets basketball team, paid about $1.5 million in 1999 for acreage in Cutchogue and Mattituck and is said to be planning a winery. Marco Borghese, an Italian with a number of silent partners, bought the Hargrave Vineyard, also in Cutchogue, for around $4 million in 1999.

Other prominent Long Island wineries sold in 1999 included Corey Creek, Laurel Lake, Peconic Bay, and Bedell Cellars. Bedell went to Michael Lynne, a Time Warner executive, for about $5 million. Gristina Vineyards was sold for about $5.5 million, Bidwell Vineyards was on the block for about $3.5 million.

With most of these sales, the founders plan to stay on making the wine while the new owners, mostly businessmen, take over management, marketing, and promotion and enjoy the prestige that comes with being in the wine business.

At the same time, many existing wineries are expanding. Bob Pellegrini, who in 1982 started Pellegrini Vineyards in Cutchogue, recently bought sixty acres in two parcels "not to increase production, but to become totally estate," he said. "Totally estate" means that he will own all his own vineyards and can call his wine estate-bottled. Pellegrini Vineyards currently buys about 40 percent of its grapes, Mr. Pellegrini said.

Buying new vineyards is not done casually. Twenty years ago, prime farmland sold for $4,000 an acre on the North Fork. Today, driven by competition among wealthy investors, the price can reach $20,000 to $30,000 an acre, Mr. Olsen-Harbich said.

Compared with, say, Pindar, Raphael is very small. But there are grape growers on the North Fork, many of them weekend farmers, whose vineyards are even smaller—far too small to support wineries. For them, Russell Hearn, Pellegrini's Australian winemaker, has created the region's first custom-crush winery in Mattituck. Called the Premium Wine Group, it produces, ages, bottles, and labels wine for these owners of small vineyards. In California similar operations, like the Napa Wine Company, produce wines for dozens of vineyard owners.

At least one longtime resident, Russell McCall of Cutchogue, has rejected all the conventional wisdom and has planted a dozen acres of pinot noir across the road from Pellegrini. Pinot noir is a temperamental grape with a high rate of failure. But then the experts said it would never succeed in California, and it has—impressively. On the North Fork, they welcome a challenge like that.

JULY, 2000

## WHAT'S NEW? WASHINGTON

Mike Hogue stopped his pickup truck and pointed across my chest out the passenger window. "Red currants," he said. "We used to have more red currants than anyone in the state but we got out of that business." Further on, he stopped again. "This used to be all peppermint; that little beat-up building there is where we distilled it. Once I went right to Chicago and sold a barrel of peppermint oil directly to the Wrigleys. We got out when that business died too." I thought: "I came three thousand miles to talk about wine and this guy's showing me a peppermint still. Something's wrong."

Nothing was wrong. I was visiting one of the most successful wine producers in the state of Washington at his farm in Prosser, and he was simply showing me how he and a lot of other people there do business. Washington State has long played second fiddle to California and even Oregon when it comes to wines. Once, that ranking may have been justified. No longer. There is an elegance, an intensity, a new degree of excellence, in the best of Washington wines, and they deserve the attention of wine drinkers. Winemaking in Washington is rarely the spiritual journey it can be in California. These folks are not retired surgeons seeking more meaningful lives, or millionaires looking for a hobby with tax write-offs. They are third- and fourth-generation farmers, hard-nosed and pragmatic; if grapes and wine don't make it, out they go. There will always be a new apple variety the Taiwanese will pay a fortune to get. And there are hops to grow, and cherries and vegetables. Mr. Hogue, the owner of Hogue Cellars, farms some 1,700 acres in Prosser, of which only 500 are used for wine grapes. Most of the grapes that go into the various wines he makes—principally merlot and cabernet sauvignon—come from other farmers whose crops are as diversified as his.

This is an unusual wine state. Almost all the vineyards are here, to the east of the Cascade Mountains, while nearly all the people live to the west. The West Side, from Seattle to the Pacific, is cool, green and rainy; the east is desertlike, hot and dry. Thanks to the old volcanoes that make up the Cascade Range, this desert is composed of rich volcanic soil in which almost anything, including wine grapes, will grow if irrigated. The melting snowcaps of those same mountains provide much of the water this bounteous land needs. In fact, water is plentiful because both the Cascades and the Rockies to the east channel their runoff into the mighty Columbia River system and its long chain of reservoir lakes.

Some of Washington's best-known wineries are on the West Side: Château Ste. Michelle, Columbia Winery, Quilceda Creek, Andrew Will, DeLille Cellars. Handsome tourist meccas nestled among the rhododendrons and evergreens of the Seattle suburbs, they offer visitors all of the wine experience except for a vineyard view. Their grapes, and often their wines, come from the farm country hundreds of miles to the east. Visitors who see only the West Side miss the essence of the Washington wine business: its vastness. It is made up of a major viticultural region, the Columbia Valley, which has within it two subregions in the southeast corner of the state, the Yakima Valley and Walla Walla. A fourth region, Puget Sound, encompasses a handful of vineyards that manage to grow wine grapes in the less-than-hospitable climate of the West Side of the mountains. There are only about 29,000 acres of wine grapes in Washington, compared with some 550,000 acres in California. But the area available for future plantings in Washington is immense. In the 10.7-million-acre Columbia Valley, only 8,000 acres are currently planted in wine grapes.

The Yakima Valley subregion encompasses 640,000 acres of which 6,000 are now in vines; Walla Walla has about 180,000 acres of which only a few hundred are in wine grapes. The Yakima Valley and Walla Walla subregions stretch from just beyond the city of Yakima, the fruit capital of Washington, eastward to the Idaho border. The Walla Walla region actually straddles the Washington-Oregon border. In fact, some of the best Walla Walla wines, including the sought-after merlots from Leonetti Cellars, are made from grapes grown over the line in Oregon. Like Leonetti and Hogue, more and more of the eastern Washington wines are becoming familiar to enthusiasts across the country. Along Interstate 82 in southeast Washington are Staton Hills in Wapato; Bonair and Covey Run in Zillah; Washington Hills Cellars in Sunnyside; Chinook Wines, just across the road from Hogue in Prosser, and Preston Wine Cellars and Gordon Brothers in Pasco. Not far from Walla Walla are Woodward Canyon and l'École 41, both in Lowden. One of the most striking Washington wineries is Columbia Crest in Paterson, about thirty miles south of Prosser. Owned by Stimson Lane, which is also the parent company of Château Ste. Michelle, Columbia Crest sits amid some 2,000 acres of vines on a ridge with a dramatic view across the Columbia River to Oregon. I wandered around the handsome reception center at Columbia Crest one day, one of six visitors in a building designed for hundreds. But then, Paterson is not exactly on any tourist route. Stimson Lane is building for the future.

Washington's wine history is brief compared with other regions. A few wines were made here in the mid-nineteenth century, but the modern wine industry dates back only to the 1960s. Then, Washington law mandated that wines and spirits be purchased only from state stores and heavy taxes were imposed on wines from out of state, discouraging any interest in good wine. Barriers began to crumble when California threatened to ban Washington apples unless Washington lowered its taxes on California wine. By 1970, there were 70 acres of chardonnay in the state, 110 of riesling, 150 of cabernet sauvignon, and 45 of chenin blanc. By 1981, there were 19 wineries in the state. In 2001 there were 160.

An early influence was Associated Vintners, a group of amateurs who experimented with winemaking in Seattle in the early 1960s. They became a commercial venture in 1962. In 1972 the company bought its first vineyard and in 1983 changed its name to Columbia Winery. David Lake, Columbia's winemaker, is probably Washington's best-known wine figure. A Canadian who was raised in England, he recrossed the Atlantic in 1977 and joined Associated Vintners in 1979.

Mr. Lake was upbeat about Washington. "We're selling all we can make, and I don't mean just the big wineries," he said. "It's taken a long time but we're beginning to get the respect we really deserve." For the moment, "selling all we can make" doesn't mean a particularly large amount of wine. Washington's production is growing but its enormous potential has hardly been tapped. It is, one might say, as large as the state itself.

JULY, 1998

# CHAPTER 8

# WHO WOULD
# HAVE THOUGHT?

There's more to wine than squeezed grapes. Corks, for instance. The best wine in the world is worthless if its cork is bad. And there have been a lot of bad corks around in recent years, thanks to the growing demand for wine in corked bottles. The bag-in-a-box is growing in popularity, but for better wine it simply won't do. What is really needed is a campaign to get wine drinkers used to metal caps. The kind used for beer and soda bottles. They work better than corks. But wine drinkers are conservative; they want corks.

Oak chips are a part of the winemaking process most people don't know about. More often than anyone might suspect, oak chips are what gives wine the taste that seems to come from months or years in oak barrels. Americans love oak, and oak chips dumped in the wine vats do all the "aging" needed in a couple of days.

Of course, not all wine is made from grapes. There is, for example, birch wine, an ancient Russian drink made in Scotland. It tastes like a good Macon blanc.

It is definitely not the kind of wine that would interest Orley Ashenfelter. Mr. Ashenfelter—Dr. Ashenfelter—who teaches economics at Princeton University, has devised his own way to predict whether a vintage, particularly in Bordeaux, will be good or bad. No sniffing, slurping, and spitting for Orley Ashenfelter. He does it all with a computer.

And then some words about wine at the White House in the early years. We know about Jefferson and Washington, but wine was popular in other presidential households as well, more popular than we have been led to think. But it was elitist. It had no recorded impact whatsoever on the rum and whiskey preferences of the common folk.


## BRITAIN'S BOAST

You are no doubt aware that, along with vodka and the samovar, the Russians invented the lightbulb, the automobile, and the airplane. At least that's what some have claimed at one time or another. But you probably didn't know that the English invented Champagne. Well, neither did anyone else, but over the years that's exactly what some articles in British publications have claimed.

Since the Middle Ages, and possibly before that, the English have been importing French wines. Four hundred years ago or so, the vineyards of the Champagne region produced quite a lot of the wine that eventually found its way to England. The wine, shipped in hogsheads and never meant to be bottled, would go through its second fermentation in the barrels after it had arrived in Britain.

As is often the case in history, different events that occur at approximately the same time, when brought together, can have a major effect on people's lives. With wine, these were the development of mass-produced glass bottles and the refinement of the cork stopper.

Vintaged Bordeaux wines date more or less from this time, around the mid-1700s. The Bordelais had known for some time that their wine was capable of improving with age, but since the wine was stored in casks, it had been almost impossible to prevent it from oxidizing. Bottles and good corks changed all that. What's more, since the Bordeaux wines went through their second fermentation while still in the casks, there was no need to worry about the bottles exploding.

In Champagne, it was a different story. There the winemakers came to realize that they could produce an excellent sparkling wine if they could determine how to keep it in the bottle. Fermentation produces alcohol and carbon dioxide. In an open vat, the carbon dioxide escapes into the air. In a Champagne bottle, it is absorbed into the wine but exerts considerable pressure on the bottle, as anyone who has ever pulled a Champagne cork knows.

Dom Pérignon, the seventeenth-century Benedictine monk who served as cellar master at the Abbey of Hautvillers, in the center of the Champagne region, is credited with solving the problem. Reportedly he not only discovered a method of controlling the fermentation but also developed both a thicker bottle and a cork stopper strong enough to withstand the pressure built up by the still-fermenting wine.

The first wine cork is said to have been developed by monks at abbeys in Spain and Portugal, where cork trees grow, and to have been carried back to France by monks or other travelers who had made the famous pilgrimage to Compostela. These corks also found their way to England, and it is this fact, rather than any claims about winemaking, on which some English historians base their assertion that Champagne was an English invention. It wasn't that they invented the wine; they were merely the first to use corks to capture the fermentation process in a bottle, beating even Dom Pérignon. Needless to say, the Champenois shrug off such revisionist theories.

"They"—meaning the English—"have been trying to claim for years that they really invented the cork," said Jean-Louis Carbonnier, of the Champagne Bureau in New York City. "I think that's what this Champagne story is all about."

In fact, French chauvinism, where Champagne is concerned, while understandable, is somewhat misplaced. Consider, for a minute, some of the major names in the Champagne business: Bollinger, Krug, Mumm, Deutz, Geldermann, Piper, Heidsieck, and Roederer. It is difficult to think of any other business so clearly influenced by one particular group of immigrant entrepreneurs.

Some of the Champagne-firm founders were born in territories that changed hands during the wars that have ravaged northern France since Roman times. William Deutz and Pierre Geldermann were born in Aix-la-Chapelle, which, for the present, is in West Germany and known as Aachen. Joseph Bollinger came from Württemberg and Johann Joseph Krug from Mainz. The original Mumm was from Rüdesheim, and

Florenz-Ludwig (later Florens-Louis) Heidsieck came from Borgholz-hausen in Westphalia.

Most of the Germanic Champagne dynasties were created in the late eighteenth and early nineteenth centuries, when Europe was relatively prosperous and destined to become more so. French prestige and power were at their peak. Moreover, it was the beginning of the commercial age—a period that marked the emergence of a new, wealthy merchant class, the people who, from one end of Europe to the other, would become enthusiastic consumers of Champagne.

Why those canny Germans were able to move in and set themselves up alongside the famous French houses—Moët & Chandon, Ruinart, Laurent-Perrier—is not clear. Many of them were from the German wine regions, and perhaps their own backgrounds gave them an objective view of the enormous growth potential for Champagne.

Little remains of the Teutonic roots of these Champagne firms, although one family at least actually returned to Germany. The Mumms, who had lived in Champagne for almost a century before World War I, found their entire business confiscated by the French government because none of them had ever bothered to take out French citizenship. The French sold the company in 1920 to a group that included the makers of the aperitif Dubonnet. The Mumms returned to Germany, where they continue to make sparkling wine in the Champagne style.

MAY, 1989

# INGENIOUS WAYS TO MIMIC THE SMELL OF OAK

Do you like a touch of wood in your wine? Do you look for that enticing taste and heady vanilla aroma of oak when you pull a cork? Do they call up an image of dark cellars and rows of barrels aging wine to the peak of perfection?

If you are in the habit of paying $20 or more a bottle, there's a reasonable chance that your wine had a barrel in its past. If you are an oak fan who finds it painful to go over $10 a bottle, welcome to the world of "alternative wood products."

There are ways to impart an oak smell and taste to wine that have nothing to do with barrels. There are oak concentrates, oak capsules, and, the most common, oak chips.

Oak chips are exactly what they sound like—bits of wood that one might find around a construction site or a lumber camp—except that they are produced commercially in large quantities. It is quite a sophisticated business: There are "French oak" chips and "American oak" chips. And since some winemakers prefer their "barrels" toasted, or slightly charred, wood chips come in varying degrees of toast.

By one industry estimate, about a million pounds of chips are sold to winemakers each year. Since from one to two pounds of chips are used for each one hundred gallons of wine, as many as 100 million gallons of wine are "aged" this way annually.

Originally barrels were simply the most efficient way to hold wine. Before bottles became widely used in the eighteenth century, barrels were the only way to keep and to transport wine. The idea that they could enhance the wine is a relatively new idea.

Today winemakers argue heatedly over the merits of not just barrels, but also the types of wood and the places where it is grown. Oak is practically the only wood used now, but chestnut and redwood were once widely employed. French oak is generally considered the best.

Winemakers who can afford to—famous chateaus in Bordeaux, some top producers in Burgundy and in California—use new barrels every year. Many put some wine in new barrels and some in older wood. Winemakers can be fanatical about the quality of the barrels they buy. Angelo Gaja, an Italian winemaker, buys the best French barrels, then has his own coopers disassemble them so the staves can age another three or four years outside his winery.

Oak adds not only flavor and aroma but also tannins, organic compounds that contribute to the aging process, particularly with white wines. In the past, some wines spent many years in wood. But the trend has been away from long aging. Now, even the prestigious Bordeaux chateaus rarely age their wines more than eighteen months, and in Italy and most of Spain, where long aging was once routine, twenty-four months in wood is considered sufficient.

Even so, contact with oak is now the norm; consumers demand it. Coopers everywhere are hard put to keep up with the demand, and prices show it. A new barrel from a famous French *tonnelier*—their word for "cooper"—can cost $600; a top-quality American barrel, $300.

To the cost of the barrels must be added the cost of the space to store them and, finally, the time the wine spends aging, when the winemaker gets no return on his costs. For these reasons, shortcuts to aging can be attractive to large-scale producers.

"If you want to make the best wine you can and cost is no object, then, generally speaking, oak barrels are used," Dr. Vernon Singleton, a University of California enologist, told *Vineyard and Winery Management,* a trade publication. "But as far as flavor extraction, oak chips and similar products can help."

Phil Burton of Barrel Builders, a Napa Valley company that sells both finished barrels and chips, said most of his customers leave chips in wine for one to three weeks, even though the chip producers say two or three days is all that's needed. The chips are usually stuffed in mesh bags and suspended in the wine barrel or tank. Often a concentrated batch of oak-flavored wine is made with chips, then blended with other wine to give the oak effect.

Mr. Burton sells both French and American chips in three toast styles: light, medium, and heavy. Buy five hundred pounds or more and Barrel Builders will custom-toast your chips. Prices range from $1.50 to $2.50 a pound for American oak and $3.50 to $4.50 a pound for French oak, depending on the size of the order. "A winemaker can add a reasonably complex oak character to a thousand gallons of wine for about $25," Mr. Burton estimated. "Chips add a significant amount of oak character at a tiny fraction of the cost of new barrels."

Mr. Burton said that his chip customers included some of California's best-known wine producers but that none wanted their names revealed. "There's nothing illegal about it," he said, "but there is always the question of image."

George Rose, a spokesman for Fetzer Vineyards in Mendocino County, said the winery used oak chips in its lower-priced Bel Arbors chardonnay. "It's strictly a question of cost," he said. "This gives us the touch of oak we want without the expense of barrel aging." Bel Arbors chardonnay usually sells for less than $10.

"Any time you find an oaky chardonnay for $10 or less, you can be pretty sure it didn't get that way from a barrel," said Rusty Eddy of Benziger Winery in Sonoma County, California. Benziger used to own the

Glen Ellen Winery, one of the country's largest producers of low-priced varietal wines. "You could say we experimented with chips at Glen Ellen," Mr. Eddy said.

A spokesman for the Robert Mondavi Winery said used barrels from higher-priced lines were used to age their less expensive lines.

Barrel Builders' chips are used mostly by wine companies at the lower end of the price scale. A company called Innerstave supplies a stainless-steel support that holds an oak stave or a number of staves inside anything from a steel drum to a thirty-thousand-gallon concrete tank. The staves work the same way as the chips, imparting oak flavor to the wine in the container. Innerstave employees chuckle when they read lists of medal winners in wine competitions. They know which ones were "aged" with their equipment.

MARCH, 1994

## GETTING A KICK FROM CHAMPAGNE

Quick: Setting out from Paris, where can one combine great art, stirring history, magnificent scenery, superb food and wine, and still be back in town in time for dinner?

Easy. In Champagne.

Ninety minutes from Paris by car, an hour by train, begins one of the most fascinating, and least familiar, sections of France. Everyone knows about *le Champagne,* probably the world's best-known wine. But *la Champagne,* the region where it's made, is something else again. A region rich in history and culture, Champagne's farms and vineyards and unspoiled villages represent the best of rural France. Its restaurants and hotels are first-quality, and its two cities, Rheims and Épernay, sit on top of several hundred million bottles of great wine.

Almost as large as Belgium, the region is nonetheless virtually unknown territory to many travelers, and particularly to the crowds going to Disneyland Paris, practically next door. Which is a pity, for Champagne is the birthplace of modern France.

The weight of history is everywhere in Champagne. The Romans came in 57 B.C. and made Durocotorum, later Rheims, an important city in Roman Gaul. The chalk vaults from which they carved their building blocks can be visited today. They shelter millions of bottles of aging wine.

According to legend, St. Remi, Bishop of Rheims, converted Clovis, King of the Franks, to Christianity in 496. In all, thirty-seven kings of France were crowned at Rheims, beginning with Charlemagne's son, Louis the Pious, in 816, and ending with Charles X in 1825. Charles VII, with Joan of Arc at his side, was crowned there in 1429.

There are no more kings, but the great French Gothic cathedral remains. Built on the foundations of a 4th-century church, it was begun in 1211 and finished, more or less, a century later. It was shelled by the Germans in the Franco-Prussian War in 1870, and again in World War I, when its magnificent stained glass windows were destroyed. Restored, partly with Rockefeller Foundation funds, the cathedral was reopened in 1938 and somehow missed the destruction suffered by much of the rest of Rheims in World War II.

Since the defeat of Attila the Hun at Chalons-sur-Marne in 451, Champagne has been a battleground. Verdun and Château-Thierry are close at hand, and at every turn in the road there seems to be another cemetery for French, German, American, British, or Canadian soldiers. In 1914, the famous *taxis de la Marne* brought up from Paris the reinforcements who helped stop the Germans at the gates of Rheims.

Épernay, the second city of Champagne, was itself burned, sacked, or pillaged at least twenty-five times in the thousand years before the seventeenth century.

Napoléon fought his last battles in Champagne. The day before Épernay fell to Russian and Prussian troops in the spring of 1814, Napoléon bestowed his own cross of the Legion of Honor on the mayor of the city, Jean-Rémy Moët, a loyal friend. Then he left for Paris and his abdication.

Despite the region's size, authorized plantings of Champagne grapes amount to only about sixty-eight thousand acres. The best vineyards are on the Montagne de Reims, between Rheims, as it is spelled in English, and Épernay, but Champagne grapes are grown as far west as Château-Thierry and as far south as Les Riceys, in the department of the Aube, southeast of Paris and only a few miles north of Burgundy.

Modern Champagne is made from three grapes: chardonnay, a white wine grape, and pinot noir and pinot meunier, both red wine grapes. The harvest is usually in late September and early October—a particularly delightful time to visit. (The cellars are open all year, but winter can be chilly and foggy.) Like all wines, Champagne undergoes a first fermentation that changes the grape juice into wine. Next comes the blending stage, when the winemaker chooses how much chardonnay, pinot noir, and pinot meunier, each of which has been fermented separately, he wants. A blanc de blancs will be all chardonnay, a blanc de noirs, all pinot noir and pinot meunier.

If he is making a nonvintage Champagne, the winemaker will add wines from as many as ten earlier years to achieve the style he wants. Wines from vineyards all over the region will be used in the blend. Few Champagne houses grow all their own grapes; some grow none. They depend on the nineteen thousand individual growers in the region.

After blending, the winemaker adds *liqueur de tirage,* a blend of sugar and yeast that starts the second fermentation. At this point the wine is bottled under a temporary metal cap.

The second fermentation produces carbon dioxide too, but since it can't get out, it remains in suspension in the wine. This fermentation also produces sediment—dead yeast cells. The bottles are placed in racks, head down, so the sediment can drift down to the neck. Each day the bottles are given a slight twist to shake the sediment down.

Next, the bottle neck is dipped into a brine solution to freeze it. The metal cap is removed and the sediment, along with a bit of frozen wine, is pushed out by the carbon dioxide. A dollop of wine and sugar, called a *dosage,* is added to make up for the frozen lump of wine and sediment just ejected, and finally the bottle is closed with a real cork.

Then it is laid down to age—at least a year for nonvintage and three for vintage. Some companies age their best wines for up to ten years. In any case, the wine is ready to drink when sold; further aging is unnecessary, although there are people who like old Champagne, with its darker color and pronounced nutty flavor.

At Moët & Chandon, in Épernay, tour guides explain all this in a brief slide show before they conduct visitors through the company's vast cellars. Moët & Chandon is the largest Champagne house, with some ninety million bottles of wine aging in its cellars at any given time, but almost every major firm and many of the smaller ones are eager to welcome visitors and reveal to them the mysteries of what is known as the *Méthode Champenoise.*

Moët is the biggest but not the oldest. That title goes to Ruinart, in Rheims, which was founded in 1729. Actually Gosset, in the village of Ay, near Épernay, dates back to 1584, but they were making still wine in those days and didn't get around to Champagne until much later.

To a casual visitor, the Champagne region is a place of prosperous farms, quaint villages, and the two substantial cities, Rheims and Épernay. There are always trucks about, laden with Champagne on its way to the world's markets, and at harvest time the smell of crushed grapes is everywhere.

But although it is big business, for the most part the Champagne business is discreetly run. Often only a brass wall plaque identifies the famous firm within. The real drama takes place out of sight, in the miles and miles of cellars that honeycomb the chalk subsoil, sometimes descending seven levels below the streets. There the bottles age in semi-darkness, high humidity, and a natural temperature that never varies from a steady 50 degrees.

The cellars at Mercier, in Épernay, are so vast that visitors are shown through them on an electric train. In the past, heads of state were shown through the Mercier cellars in horse-drawn carriages. Mercier is the largest-selling brand of Champagne in France and consequently the most visited house. Each year it receives a quarter of a million guests in its cellars in the Avenue de Champagne. The company is owned by the Groupe Louis Vuitton Moët Hennessy, L.V.M.H., whose other Champagne holdings include Moët & Chandon, Veuve Clicquot-Ponsardin, Pommery, Canard-Duchêne, de Venoge, and Ruinart.

Épernay producers that can be visited only by appointment include Pol Roger, Churchill's Champagne house of choice, and in nearby Ay, Bollinger and Deutz. Appointments to see cellars are not hard to make. A fax or a phone call will do it; or ask a hotel concierge or the Champagne tourist offices in Rheims and Épernay. Language is never a problem; even the smallest company has someone around who speaks English. When I was there, a young English guide at Moët noted rather slyly that Mercier had recently added a guide who understood the Scots.

Someone once decribed Rheims as a city perched atop a giant Gruyère cheese. If so, it's a valuable cheese indeed. Mumm alone keeps 35 million bottles of its Cordon Rouge and other labels under its headquarters in the Rue du Champ-de-Mars. Along with a tour of the cellars, Mumm offers a film and commentary on Champagne production, a visit to its museum, and a tasting.

Pommery's eleven miles of wine cellars include spectacular galleries carved out of the chalk by the Romans. The Taittinger tour offers a tasting and a visit to the *caves,* the vestiges of a 13th-century chapel, and the crypt of the 15th-century abbey of St. Nicaise, all on the property. Veuve Clicquot–Ponsardin's tour and tasting are by appointment. Charles Heidsieck, Ruinart, Henriot, and Roederer are other Rheims-based Champagne houses that welcome visitors only by appointment.

Most visitors head for the big companies whose names they know, but for French-speakers, a visit to one of the many small producers out in the Champagne villages can be even more rewarding. Customers drive out from Paris, share a fraternal bottle with the proprietor at his kitchen table, buy a few cases, and haul them home in the trunk of the car. Several hundred of these small houses do a thriving "cellar door" and mail-order business.

My favorite is the house of Ricciuti-Révolte, in Avenay-Val-d'Or, near Épernay. Baltimore-born Al Ricciuti met his wife-to-be while serving in the army in France during World War II. Later he moved to France and joined her family's Champagne business, which he now runs. He is undoubtedly the best Champagne maker ever to come out of Baltimore.

Another inducement to visit the region is the restaurant renaissance that has taken place there, led by Gérard Boyer's elegant and expensive Les Crayères, just outside Rheims. Among the best are the old, reliable Royal Champagne at Champillon, on the road between Rheims and Épernay; Au Petit Comptoir, in Rheims; and the Château de Fère at Fère-en-Tardenois.

Épernay and Rheims both consider themselves the capital of Champagne, and if possible a visitor should experience them both. For train travelers, Épernay might have an edge, as the Avenue de Champagne with nine famous houses is only a few steps from the railroad station. Some hardy voyagers board another train at Épernay and go on to Rheims. Bottom line? Rent a car in Paris and drive out. Or better yet, take the train and rent a car in Épernay. Tour the region at leisure, return the car to Épernay, and relax on the train back to Paris. Tell the conductor you've just spent an unforgettable day or two in Champagne and to please wake you at the Gare de l'Est.

The Benedictine monk Dom Pérignon, administrator and cellar master at the Abbey of Hautvillers from 1668 to 1715, is wrongly credited with "inventing" Champagne by tying down the corks on his bottles and preserving the fizz.

In fact, he took every precaution to prevent his wines from fermenting in the bottle and was known in his own time for his skill at winemaking and in selling his wines. He was, apparently, a teetotaler. He can be seen, in effigy, at his desk in the abbey, which is just over the Marne River from Épernay. The ruins of Hautvillers were purchased in 1822 by Comte Pierre-Gabriel Chandon de Brialles, a son-in-law of Jean-Rémy Moët. But it wasn't until 1936, when they established the first luxury brand of Champagne, that Moët & Chandon used the monk's name. They called it, of course, Dom Pérignon.

Actually, the wines of Champagne "fizzed" long before Dom Pérignon was born. Wine made in the fall, just after the harvest, will continue to ferment until cold weather stops it. Then, in the spring, it begins again. (Fermentation is the process by which yeast attacks the sugar in the grape juice and turns it into alcohol and carbon dioxide.) In the days before wine was bottled, fermentation recommenced harmlessly in the barrels where the wine had lain through the winter. The carbon dioxide escaped into the air. When bottling began in the eighteenth century, the restarted fermentation caused the bottles to explode. Which is why bubbles were the last thing Dom Pérignon wanted to see in his bottled wine.

SEPTEMBER, 1996

## WINE MADE FROM OTHER THAN GRAPES

A casual glance at the label and you might conclude that there is yet another new winery in California, one called Silver Birch. Well, take another look. This wine is made from the silver birch tree, and it is produced in Inverness, Scotland, at what is, so far as anyone can determine, the world's northernmost winery.

People who make wine from something besides grapes often feel that the world has forgotten them. One can hardly blame them. Shelves are filled with wine books that never even acknowledge that wine is not necessarily made from grapes.

Then there is the image problem: genteel pensioners sipping dandelion wine, the bottle of cherry wine brought out for the minister's annual visit, that sort of thing. But grapes will not grow everywhere, and the fact is that other fruits and plants can be turned into pretty good wine.

It is a question of natural sugar and acids. Grapes have ideal quantities of both to make fine wine; quite a few other fruits and a few plants do too. And if they do not, well, at the schools of enology they call it "adjusting." As in "We adjusted for sugar content," which means a few sacks of sugar were added to the wine during fermentation to get the alcohol level up a point or two.

Some years ago, a New York banker named Frank Jedlicka got into the apple wine business in Vermont. On the side, for his own amusement, he made wine from just about everything around his place except the cuttings from the front lawn. On one memorable occasion, he asked a group of French guests to try to identify the fresh dry white wine he had served with dinner. Muscadet was mentioned, probably Sancerre and a couple of other things. He had made it from celery. The French, he recalled, chuckling, were not amused.

Which brings us back to the silver birch and the wine made therefrom. According to one report, Queen Victoria drank it regularly at Balmoral Castle, a summer home for British royalty an hour or so south of Inverness. Another story has her sampling it. From what we know of Victoria, sampling seems more likely—she was not one for overindulging. But in spite of her Hanoverian roots, she took on, however slightly, the coloring of a Stuart when she summered on the Speyside. And what more appropriate than to partake of the local fare?

If what the Queen tasted was anything like what is produced here today, she enjoyed a totally dry, well-balanced white wine with strong kinship to wines made from the sauvignon blanc grape. The taste is French and the body more Californian. One would have to be an expert indeed to guess that the stuff came from a tree.

Not that wine from trees is totally unknown. If memory serves, Mr. Jedlicka made a maple wine, and there are home winemakers in New England who do too. And the Russians have made wine from the birches in their trackless forests for a thousand years.

The source of this oddity in Scotland is Moniack Castle, at Kirkhill, a seat of the clan Fraser for more than four centuries, seven miles west of Inverness between Loch Ness and Beauly Firth. The current chatelaine of

Moniack, Philippa Fraser, began making wine as a hobby. In 1980 she went commercial, producing four wines, a mead, and a liqueur from sloe berries and gin. The wines, in addition to the silver birch, are a dessert wine called Meadowsweet, made from herbs; an elder flower, made not from elderberries but from the flowers of the plant; and a blaeberry—blueberry to those of us from more southerly latitudes—which grows on tiny bushes on the Highland grouse moors. The blaeberry is Moniack Castle's only red wine.

The silver birch is the most important of the lot; Mrs. Fraser makes from twelve thousand to fifteen thousand bottles of it in a good year. It comes from the sap of the birch trees. Compared with grape picking, which takes on many of the aspects of a military operation each fall, sap gathering is relaxed and uncomplicated. A hole is bored in the tree "about the size of an old English penny," Mrs. Fraser said. A cork is placed in the hole and one end of plastic tube, a quarter-inch in diameter, is inserted into the cork. The other end goes into a five-gallon plastic drum.

The sap begins to rise around mid-March, often when snow is still on the ground. Between then and mid-April some three hundred trees are tapped. Each produces about five gallons of sap. Thereafter, the winemaking process is much the same as it is for grapes. The wine is allowed to mature for more than a year before it is released, but it shows up in the castle's own shop and in restaurants around Scotland well before the traditional grape wines of the same vintage because the sap is gathered six months earlier than the same year's grapes.

Needless to say, Scottish wine is somewhat less well known than Scottish whisky. But after visiting and tasting at a few single-malt distilleries, Mrs. Fraser's little winery is a welcome respite. There is also a restaurant, where a lobster and a bottle of silver birch in front of the roaring fire can help ward off the dampness and the cold.

And both of those are endemic to the region. *Moniack* means "little bog" in Gaelic, and the castle, built in 1580, stands on the first bit of dry land above sea level at the entrance to Beauly Firth. It was originally a garrison to defend Fraser lands, and became a family home in the eighteenth century. Mrs. Fraser is the widow of the late Major A. H. Fraser, a first cousin of Lord Lovat, the chief of the Fraser clan.

Mrs. Fraser started the winery, she said, to avoid having to sell Moniack. Besides the wines and liqueurs and the restaurant, which serves lunch and dinner, there is a line of unusual homemade preserves. They include a sharp rowanberry jelly meant to complement venison and a smoky hawthorn jelly to go with game birds and pork.

SEPTEMBER, 1989

## A FORMULA TO RATE BORDEAUX

The wine on the table was Château Bois de Notre Dame, 1970, an Haut-Médoc. Never heard of it? Neither had I. Neither had Orley Ashenfelter, who had brought it to the noisy little West Village restaurant.

"Hundred bucks a case at an auction in San Francisco," Mr. Ashenfelter said, chuckling. It was delicious, and at $8 and change for a good twenty-year-old wine, he was entitled to be a bit smug.

But that wasn't why we were there. The idea was to talk about the Ashenfelter Theorem, sometimes referred to as the "Bordeaux equation." Mr. Ashenfelter teaches economics at Princeton, a calling that, to hear him tell it, paid just below a subsistence wage.

Combining a marginal wage and a fondness for claret is no easy task. Mr. Ashenfelter discovered early on that the place to go was the auction hall. Most stories about wine auctions dwell on owners of fast-food franchises who snap up ancient Lafites at $20,000 a bottle. Hardly anyone knows about the genuine bargains that regularly go on the block. Orley Ashenfelter knows. Witness the Bois de Notre Dame. As a wine lover, he had stumbled on a good thing. As an economist, he couldn't leave well enough alone: He had to turn it all into a formula. It was that formula, or equation, that sent a *frisson* through the wine world when it was outlined in an article in *The New York Times* in 1990.

What Professor Ashenfelter has done is this: create a formula based on winter rainfall and growing-season temperatures in the Bordeaux wine region that, he says, enables him to predict the quality of the annual wine vintages. For some reason, this rather obvious bit of cerebration has prompted an astonishing outflow of consternation, not to mention indignation, in the wine establishment.

To be sure, Mr. Ashenfelter has fine-tuned his formula and spiced it up with some forbidding jargon. But the fundamental relationship of which he writes—the connection between weather and wine quality—is well known to any fourth-grader in the Bordeaux schools.

By applying his formula to the heavy winter rains and exceptionally hot summer of 1989, Mr. Ashenfelter concluded that the 1989 Bordeaux vintage could be the finest of the century. Well, great, but he was hardly going out on a limb. Everyone in France had been raving about 1989, most of them since the fall, when the grapes came in.

So why all the fuss? Two reasons. Some elements of the wine trade are angry because the Ashenfelter equation could be helpful in identifying and steering customers away from lesser vintages they have promoted. Second, and more seriously, he is accused of relegating the whole wine-tasting mystique to a minor role. Supposedly the sipping, spitting, sniffing, and note-taking so dear to wine romantics have all been rendered obsolete by mathematics.

The pessimists are wrong on both counts. In the first place, Mr. Ashenfelter is not doing anything new; he's refining techniques that winegrowers have been using for centuries. And the tasters are far from obsolete. Mr. Ashenfelter can tell you a certain vintage is going to be good. But there are going to be some wines that year that are better than others and his formula can't identify them. He can argue that a year is poor, but there may be some exceptions and his formula can't single them out.

The professional tasters—those who buy from the chateaus—begin their work in the early spring after the harvest. They already know what the overall quality of the vintage is; their work—deciding which wines to buy, which to reject—is with the individual chateaus.

So too the critics, like Robert M. Parker Jr., whose numerical ratings virtually dictate what fine wines are sold in this country. If someone—Mr. Ashenfelter, for example—can help everyone get a better fix on the individual vintages, Mr. Parker's ratings will be more valuable than ever. A vintage assessment is merely another tool in judging wine. Since there are good wines in poor years and vice versa, the job of the traditional critic will always be important.

The people who worry most about vintages are those who invest in wine futures. They buy early in the expectation that the value of their wine will go up. An early assessment by Mr. Ashenfelter might influence people's decisions to get into the market or not. A few people invest in Burgundy and California, but most wine futures business is in Bordeaux.

And what about the good Bordelais? Will a word or a number from Or-ley Ashenfelter consign whole vintages to the distilleries? Not too likely; much of Mr. Ashenfelter's work was originally done by Bruno Prats, the former owner of Château Cos d'Estournel. Mr. Prats has been making cal-culations like Mr. Ashenfelter's, though admittedly not as precise, for years. So far, his effect on the market has been minimal.

Which brings us back to that bottle of Bois de Notre Dame. Mr. Ashen-felter gives 1970 a quality rating of 43. The control is 1961, with a rating of 100. It was really good, even if neither of us could locate the chateau.

MAY, 1990

# 1993: ANOTHER GREAT YEAR FOR PARIS'S VINTAGE

The 1993 vintage has not generally been much of a success in France. Bad weather has caused problems almost everywhere. Except in Paris, where every vintage is a success.

Most Parisians and many visitors know about the little vineyard up on the slope of Montmartre, near the white-domed basilica of Sacre-Coeur; some know about the even smaller patch of vines in the Parc Georges-Brassens in the 14th arrondissement, in the southern part of the city; and quite a few are aware of the huge—and quite fruitful—vine that Jacques Melac has trained around the outside of his wine bar, Les Caves Melac, on the Rue Léon-Frot, not far from the Place de la Nation, in the 11th arrondissement.

But hardly anyone knows about the dedicated Parisians from Neuilly to Vincennes who raise grapevines in backyards, on terraces, even in window boxes. Even the British Embassy is said to have an inconspicuous little patch of vines somewhere in its garden.

The French are demon gardeners, and until recently most of these would-be winemakers nurtured their charges alone, without the comfort that comes from shared hopes and frustrations. All that has changed, thanks to a group called l'Association des Vignerons de Paris, "the Paris Winegrowers Association," founded in 1991. "We have a hundred growers, including the British Ambassador, and another one hundred associate members who sup-port our goals," said Mr. Melac, one of the founders.

Each year on the first weekend of October, members riding two gaudily decorated *charrettes,* the old-fashioned horse-drawn farm wagons, and accompanied by the Postal Workers' Band traverse the city and stop at well-known street markets to enlist new members in the association by distributing free vines. They give away about ten thousand shoots, most of them a hardy version of the pinot noir grape donated by a plant nursery in Champagne, along with a booklet explaining how to grow them.

On Sunday morning, after the last vines had been distributed, members of the association showed up with the grapes at the *mairie,* or city hall, of the 11th arrondissement, not far from Mr. Melac's wine bar, to take part in the group's second harvest. Slightly more than a ton of grapes are collected. Some families load tubs filled with grapes into station wagons, while others bring their small harvests on the Métro.

The grapes are crushed by students, also vine growers, from a local school. They also make the wine, under the watchful eye of Philippe Dhalluin, the winemaker at Château Branaire-Ducru, in Bordeaux. The wine is stored in casks in one of the remaining old wine warehouses at Bercy on the eastern edge of Paris. The 1993 harvest produced about one thousand bottles, Mr. Melac said.

The association's first vintage, 1992, brought in about 1,650 pounds of grapes, enough for six hundred bottles of wine, some of which were auctioned off for charity at Bercy-Expo, the city's new convention and exhibition center. The rest were distributed to the grape growers.

In the Middle Ages, the Paris region was one vast vineyard. Monasteries cultivated the land for miles around the relatively small city, providing for virtually all its agricultural needs, including wine.

As late as the French Revolution, the region had some forty thousand acres of vineyards. In the first half of the nineteenth century, people in little towns like Passy and Auteuil, now just Paris neighborhoods, grew vines much as people do in villages in Burgundy today.

In the 1870s, when a phylloxera infestation devastated the vineyards of France, it seemed at first that the Paris region would be spared. The local vintners made fortunes pushing their thin, acidic wines until finally the disease hit them too.

By the time the vineyards had recovered, about twenty years later, new methods of transport had transformed the face of France and Parisians could buy rich, full wines from the south that cost far less than the unattractive local wines. Industrialization and urbanization soon finished off the rest of the commercial vineyards in the Paris region.

In 1933, prodded by local artists, Montmartre got permission from the city of Paris to plant a small vineyard on empty land near the Sacre-Coeur. The vines are never particularly prolific, but each year Montmartre celebrates the harvest with a day of festivities, usually led by film stars, bands, majorettes, members of wine fraternities in their mock-medieval robes, and of course politicians mindful of the television coverage.

After the harvest, everyone gathers in the basement of the local *mairie* to watch the symbolic crushing of a bunch of the new grapes and to share a glass of the previous vintage. Then the rest of the year-old wine is auctioned off for charities.

Mr. Melac, who is almost as good at attracting publicity as he is at running one of Paris's best-known wine bars, has been holding his own harvest every year for many years. His single enormous vine, a baco noir, climbs the side of his building and grows along a ledge around the front and side.

Each autumn, while a band provides noisy inspiration, Mr. Melac and his friends mount ladders to gather the grapes. The grapes are crushed by barefoot girls and the juice carted off to be turned into, well, some kind of wine.

The work done, the harvesters and a crowd of Melac regulars turn to wines decidedly not from Mr. Melac's vine and to serious quantities of *charcuterie* and cheeses from his native Auvergne, all set out on tables in a blocked-off side street.

It's too early to tell whether the harvests at Montmartre and the Caves Melac will combine with the activities of the winegrowers association. "Why have fewer wine celebrations?" Mr. Melac said. "We should have more."

In fact, another harvest festival is in the making. Inspired by the success of the association, Pierre-Henri Taittinger, mayor of the 16th arrondissement and a member of one of the great families of Champagne, last year planted some of his family's vines in the Trocadéro Gardens in his district and announced, only slightly tongue-in-cheek, that in a few years he plans to produce, and celebrate, the first Paris Champagne.

OCTOBER, 1993

## THE  TALE  OF  A  GRAPE

V ouvray is one of those wines that stir up strong feelings of *ennui.*
There are problems. First of all, the grape: chenin blanc. Chardonnay it's not. "A magical chameleon of a grape," says Jancis Robinson in her valuable book, *Vines, Grapes and Wines.* But, she adds, "Remarkably few of the thousands of growers who harvest it each year even realize that magic is there."

Could have fooled me, too. Two decades ago, chenin blanc was big in California. But with one or two exceptions, it never rose above mediocrity; the wines were eminently forgettable. (Not including Chappellet's dry version.)

I was thinking about Vouvray, for the first time in years, aboard the spectacular TGV (*train à grande vitesse*) Atlantique, en route from Paris to Tours. The high-speed train slices right through Vouvray just before it slows down to stop at St. Pierre des Corps, the main-line stop for Tours. And thereby hangs a tale.

Back in the early 1980s, when the French National Railroad was planning the western line of the TGV, it became clear that the new tracks would destroy acres of valuable Vouvray vineyards. The growers rose up in protest, blocking the traditional main railroad line from Paris through Tours to Bordeaux and vowing to halt any work on the new line that might bruise a single Vouvray grape.

In France one doesn't brush off the railroad unions, particularly if one is a Socialist president. Here was a serious impasse. Translation: a good story.

I quickly found myself in Vouvray, which is little more than a wide spot in the road that runs along the north side of the Loire River three or four miles east of Tours. The growers were angry; the minister of agriculture had promised them, like Pétain at Verdun in 1914, that the train would never pass through their vineyards.

Fine, said the railroad people; your vineyards are on top of a cliff. We'll tunnel right under your dumb grapes.

Back to the minister went the growers. "Fellows, I'd like to help you," that statesman said in effect, "but my portfolio covers the topsoil and what grows in it. Tunnels are out of my hands."

The railroad, it seemed, had won. Or so the bitter growers had concluded. After an hour or so with them, it was time to interview the mayor, a fellow named Gaston Huët, who happened also to be one of the town's most important winemakers.

"First, it's time for a glass." he said, producing a squat bottle of golden-hued liquid, a 1945 from his Le Haut-Lieu vineyard. Mr. Huët had been around a long time, as mayor and winemaker. Wine and politics both need time. One didn't have to be a historian to know what straits France had been in in 1945. Was this '45 bottle meant to be some kind of metaphor for the local crisis? It was quite a wine—subtle, many-faceted, still lush and complex after forty years in the bottle. Rabelais, who once lived in the neighborhood, described Vouvray as being "like taffeta." Until this bottle, I'd always thought that to be typical Rabelaisian hype.

We tend to prattle on about dry wines in this country, even though, as many a cynical winemaker will readily acknowledge, we are happiest with a bit of sugar, residual or other, particularly in white wines. The Huët Vouvrays are *demi-sec,* half-dry. Translation: slightly sweet, but only slightly. There is an acid balance in the best of these wines that Sauternes might envy.

Even so, there is a problem of placement. The wine is not dry enough, it's thought, for a table wine, but too dry for a dessert wine. Be assured: It can be both.

Simon Loftus, a British writer and wine merchant, put it this way: "Loving the wine but slightly skeptical of how to drink it, I took a fine bottle of demi-sec Vouvray home with me and tried it with salmon trout. It was a perfect marriage. I tried it with cold chicken; it was superb. I tried it by itself and it remained irresistible." Mr. Loftus noted, too, that he had drunk fifty- and sixty-year-old Vouvrays that were still youthful and fresh.

Gerald Asher, writing in *Gourmet* magazine, fondly recalled an 1895 Vouvray drunk not too many years ago whose bouquet and flavor were "honeyed and rich, with suggestions of impossible tropical fruits."

For a period beginning in the 1950s, most of the wine to bear the ancient name of Vouvray was cheap sparkling stuff, destined for grocery chains and for nightclubs around the Place Pigalle. By the early 1970s, the market was glutted with banal sparkling wines (as it still is) and the Vouvray producers decided to go back to their origins. One sign that things were beginning to come their way: Patrick Ladoucette, perhaps the most savvy marketer in the Loire Valley, bought the house of Marc Bredif, traditionally one of the most respected of the Vouvray producers.

Today, Bredif, Gaston Huët, along with his son-in-law Noël Pinguet, and André Foreau are three of the top names in Vouvray. All produce the classic demi-sec wine and, when weather conditions permit, a concentrated sweet wine that rivals the *vins liquoreux* made anywhere else in the world, whether it be Sauternes, the Rheingau, the hills of Alsace, or the Napa Valley.

Oh yes, the train. It did indeed go through a tunnel under the vines. But, urged by Mr. Huët, the railroad agreed to lay thick rubber mats under the tracks to soften the vibration. Now the vines grow undisturbed and the growers, should they desire, can zip off to Paris on the TGV in the time it took their fathers to get downriver to Tours, a few miles away.

JUNE, 1991

## LESS WINE, MORE WATER

In 1990, a study showed that the French people are turning away from wine and that their favorite replacement was water. In fact, the study showed that for the first time anyone can recall, a majority of French adults say they never drink wine.

The countrywide survey was undertaken by the National Interprofessional Office of Wine, a promotional arm of the industry. The group reported that 51 percent of the adult population drinks no wine at all. In a 1980 survey by the same office, 39 percent of the respondents said they never drank wine. By 1985 that figure had climbed to 45 percent.

Some 4,000 people were queried about the drinking habits of all the adults in their households, a total of 12,400 people, the wine office said.

Still, 28 percent of the men interviewed said they were daily wine drinkers. Only 11 percent of the women drank wine daily. The principal reason for abstaining: 73 percent of those who said they do not drink wine said they don't like the taste. About 25 percent spoke of health concerns.

While the 51 percent figure is noteworthy, it is only another landmark in the long-term decline of wine drinking in France. And, for that matter, in Italy and Spain. In all three countries wine consumption today is considerably less than it was at the end of World War II. It is probably safe to say that the term "wine-drinking nation" may be obsolete early in the next century.

In 1930, per capita consumption of wine in France was about 130 liters (almost 35 gallons) a year. By 1950 that had dropped to 109 liters, and by 1986, according to figures compiled by the European Economic Community, to about 78 liters. Spain and Italy saw increases in consumption during the late 1960s and '70s, but in both countries consumption declined considerably in the 1980s, in both cases to levels below those of 1930 and 1950.

In the E.E.C., wine is the fourth most popular drink after tea, coffee, and beer, in that order. Wine and soft drinks are about even, and mineral waters are sixth.

In *Wine in the European Community*, a 1988 study, the E.E.C. listed five general factors as possible causes for the decline in wine consumption in traditional wine-producing nations, in addition to "excessively high taxation and serious economic depression."

- The drop in rural population as people move to the towns.
- Changes in dietary habits resulting mainly from changes in the pattern of working life.
- Increased consumption of other beverages, as a result of extensive advertising.
- Misleading information regarding the authenticity of wine or its effect on health.
- A greater propensity of consumers to spend their money on other things.

Some of the E.E.C.'s reasoning is questionable. Europeans have been moving off the land in large numbers for almost a century. Even so, as late as the 1950s, cities like Rennes in France reported that in some neighborhoods fewer than 20 percent of the domiciles had running water. Those people continued to drink wine. Certainly changing work patterns played a role. In the early part of this century it was not uncommon for laboring men—factory workers, farmhands—to drink 10 liters of wine a day. And they consumed enormous meals; the legendary *trou normand* was a "hole" made by consuming a glass of fiery Calvados after the first five courses of a meal.

The "hole" made space for the next five courses. Needless to say, enormous amounts of wine were consumed with such meals.

Just as food has lightened up—nouvelle cuisine was a response to new eating habits—wine has become less alcoholic, lighter-bodied, easier to drink.

While advertising certainly plays a role in influencing consumer purchases, it is likely that drinking habits would have changed in any case. More than anything else, the availability of clean running water in even the poorest cities and towns lessened the need for wine as a safe beverage. So did the ever-growing availabilty of wholesome food; in the absence of anything better, wine had been drunk for its nutritive value too.

The coming of electricity to the smallest hamlets made refrigeration a reality. With it came soft drinks, bottled juices, and of course beer. Invariably, the new products tasted better than the cheap wine that had been the staple of the French working class.

The gradual decline in wine consumption in the Common Market has imposed severe economic burdens on the countries involved. At present, more than 20 percent of the wine produced in France each year is turned into pure alcohol. The government, prodded by a powerful agricultural lobby, buys surplus wine from the producers, distills it, and stores the alcohol. According to E.E.C. figures, France's stocks of wine alcohol rose from some 250 million gallons in the early 1970s to almost half a billion gallons in 1986. Storage costs alone run into the millions of dollars each year.

In France, as in the United States, the trend has been to drink less wine but much better wine. To that end, the government has for twenty years been encouraging growers in the Midi, the hot shore of the Mediterranean, to pull up their cheap, heavy-producing vines to plant vines that yield fewer but better-respected grapes, like cabernet sauvignon and chardonnay.

The number of new French wineries producing top-quality wines in what were once enological backwaters is remarkable. The government has responded by upgrading such grape-growing areas, like Minervois and Côtes de Provence, by awarding them the coveted Appellation d'Origine Contrôlée. This top rating assures the consumer that certain quality standards will be maintained, and virtually guarantees the grower and winemaker a much higher return on his investment.

The proof is in any French supermarket. Once the shelves were lined with plastic bottles of anonymous cheap red wine, sold by alcoholic content (a liter at 11.5 percent alcohol cost a few centimes more than a bottle at 11 percent, and so on). Now the displays feature famous Bordeaux chateaus, the best Burgundies and Champagnes, and, more often than not, a modest display of what are called here *vins de Californie.* And if that is not a change in French drinking habits, then what is?

AUGUST, 1990

## WINE AT THE WHITE HOUSE

When President Clinton was toasted in 1993 at the luncheon of the Joint Congressional Inaugural Committee, it was with sparkling wine from Korbel Champagne Cellars of Sonoma County, California. The table wines at the lunch were from another California winery, Jekel Vineyards, in Monterey County.

Wine has been part of Washington festivities since the first President took office, but with the exception of Thomas Jefferson, few chief executives have indicated any particular enthusiasm for it. Jefferson's fascination with wine, including his efforts to grow wine grapes in Virginia, have been well documented.

During his tenure as ambassador to France, from 1785 to 1790, Jefferson took great care to develop a good cellar, and when his diplomatic posting ended, he arranged to have wine shipped back to the United States, both for himself and for his boss, George Washington. Once in the White House, Jefferson left the ordering of whiskey, beer, and cider to the staff but attended to the wine purchases himself, occasionally consulting with his French majordomo, Joseph Rapin.

When he moved to the White House in the spring of 1801, Jefferson brought Rapin with him. But Rapin, seeking a better job, only came to fill in until someone else could be found. Later that year, Jefferson engaged Étienne Lemaire, who was to run his household through two administrations.

Lemaire barely spoke English; he kept all his accounts in French, and when it came time to write his memoirs of his White House years, he wrote them in French too. He was paid directly by Jefferson—there was no budget for servants or, indeed, for any household expenses.

Lemaire notes that one of his most important duties was holding expenses down, including keeping an eye on wine consumption. Jefferson bought some bottled wine by the case, but more often it arrived in barrels. They were stored under lock and key, first in the White House basement and later in a wine cellar that Jefferson had dug on the grounds. (The first Presidential wine cellar also served as an icehouse and was usually referred to by that more democratic name. A contemporary report described it as round, about sixteen feet deep, lined with bricks and topped with a wooden shed.)

Though he had a French majordomo and, for a time at least, a French chef, Honoré Julien, Jefferson had no separate wine steward. Nor did he permit servants to remain in his dining room, except at major functions. In keeping with his egalitarian principles, guests served themselves at his table. Servants brought in the food and wine, placed them by the table, then left. Guests helped themselves and one another.

The most famous inaugural celebration in the early history of the White House was Andrew Jackson's, on March 4, 1829, when a horde of boisterous Jackson supporters descended on the mansion. The impression that has come down through history is that the mob was unexpected. In fact, the White House staff had been preparing for a big crowd for days, working quietly because John Quincy Adams and his family were staying in the house until the day before the inauguration.

The task fell to the Adams's steward, Antoine Michel Giusta, another Frenchman. His work was complicated by the fact that Jackson, who was staying nearby at Gadsby's Hotel, refused to communicate with the outgoing President. Jackson believed that the bitter election campaign had contributed to his wife's death, and he blamed Adams.

Huge quantities of lemonade and orange punch, liberally laced with whiskey, were prepared in advance and stored in the icehouse. After the swearing-in at the Capitol, Jackson rode his horse to the White House, followed by the crowd. The staff expected large numbers of people to pass through the house and stop briefly to shake hands with the President. But they didn't leave.

Seeing the President being crushed against a wall and gasping for air, several aides dragged him from the room and out of the building. To draw the crowds—described by one reporter as "a rabble, a mob, scrambling, fighting, romping"—out of the house, Giusta had the servants drag washtubs of whiskey punch out onto the White House lawn. At what time the punch ran out is not recorded.

Jackson has come to be associated with the raw-boned whiskey-drinking men of the frontier, but he observed the amenities at the White House. When Giusta quit in 1834 to open an oyster bar in Washington, he was replaced by Joseph Boulanger, a Belgian. Hired as a chef, he took over most of a steward's duties and brought back good cooking and wine.

By then the wine cellar had been brought into the White House. It was just beneath the State Dining Room. According to one report, racks for bottles and barrels were built along the walls behind heavy wooden bars, and there was room not only for wine but also for hard liquor and beer. An account of a small dinner for five in 1834 indicated that Sherry, Madeira, and Champagne were poured constantly throughout the meal and for a considerable time thereafter.

James Polk's wife, Sarah, has often been accused of banning alcohol in the White House, but she did not. A dour woman, she did ban dancing and most other forms of innocent enjoyment. But while she hated drinking, she did not forbid it, even though the temperance movement was strong in the 1840s. She banned whiskey but increased the use of table wine. As William Seale noted in his book *The President's House,* this was no economy measure. In the mid-19th century, whiskey was an inexpensive commodity, much cheaper than wine.

JANUARY, 1993


# THROUGH BORDEAUX, FROM B TO X

**Agassac, Angélus, Angludet, Ausone . . .**
"Back in the mid-1960s, my wife and I took up Bordeaux reds." Thus began a June 12, 1991, letter from a Connecticut reader that outlined a most unusual wine odyssey. The letter went on: "Early on, we began a tradition of having a red Bordeaux chateau every Saturday at noon, choosing food (even pizzas) that tended to work well with a Bordeaux red. We've been so regular in this custom that by now we've enjoyed over six hundred chateaus."

**Beau-Site, Belair, Beychevelle, Brane-Cantenac . . .**
Six hundred Bordeaux chateaus! How many Bordeaux chateau owners

could match that: their own and 599 others? How many enologists, wine-makers, growers, merchants, or wine writers could make that claim?

The two heroic Bordeaux fans are Robert L. and Frances McLaughlin, of Cheshire, Connecticut, about halfway between New Haven and Hartford. He is an economic forecaster who publishes *Turning Points,* a newsletter; she, a high-school English teacher.

### Cabannieux, Calon-Ségur, Cheval Blanc, Clerc-Milon . . .

"I was strictly a six-pack beer man in my younger days," Mr. McLaughlin said in a telephone interview. "It was after I got my M.A. at Syracuse and was married that I first got into wine. Some of our Syracuse friends were teachers, and I remember them always bringing out the Sherry. One fellow surprised us with a bottle of red wine. I'll never forget it; it was from Almaden.

"Once, on a trip to New York, I walked into the Sherry-Lehmann wine store. I told the clerk I liked the idea of trying something French. He recommended Château Gruaud-Larose. I brought it home and we had it that Saturday and it was better than a chocolate malted. That was my first real introduction to great wine. Oh yes, the Gruaud was a 1964."

### Dassault, Dauzac, Desmirail, Doisy-Daëne . . .

The Gruaud-Larose should have begun a beautiful friendship, but it wasn't that easy. The McLaughlins were disappointed in many wines they tried in the early days. "We were in our early thirties," he recalled, "and we were buying mostly cheap stuff. I can remember saying to my wife that de Gaulle must have been sending us all the junk. We'd never even heard of tannin." (Tannins are astringent substances that come from the skins and seeds of grapes and give structure and longevity to the best wines. When the wines are young, they can taste harsh.)

### Église, Enclos, Escadre, Évangile . . .

The McLaughlins have turned their Bordeaux drinking into a pleasant ritual. Saturday lunch—outdoors in summer—is planned around that week's chateau. While drinking the week's selection, they pore over their collection of Bordeaux books, reading about the wine and comparing it with others they recall from the same region or commune.

"With Féret, Johnson, Parker, Peppercorn, Penning-Rowsell, and

Hubrecht Duijker, you can really have fun on a Saturday, studying the one chateau you're drinking that day," Mr. McLaughlin said, referring to some of the better texts.

Their Bordeaux bible is the huge 1,867-page *Bordeaux and Its Wines,* published in Bordeaux by Féret et Fils and known throughout the wine world simply as *Féret.* Now in its thirteenth edition, *Féret* can be traced back to the first major Bordeaux reference work, *Bordeaux, Its Wines and the Claret Country,* published in 1846 in London by Charles Cox. "We pick our wines out of *Féret,*" Mr. McLaughlin said, "then buy them at our favorite wineshop in Hamden, Bob and Ben's."

Bob and Ben's is more formally known as Mount Carmel Wines and Liquors, but not to the McLaughlins. "Bob and Ben's is a fantastic place," he said. "They have a magnificent selection." Often, when his choice from *Féret* is not available, Bob McLaughlin lets himself be steered to something that is, like the 1987, and later the 1988, Château La Lagune.

### Figeac, Filhot, Fourcas-Hosten, Franc-Mayne . . .

As they forge ahead through *Féret,* the McLaughlins still find time to revisit old favorites. "I'd say we're on a five-year cycle with the classified growths," he said. The classified growths are the sixty-one wines selected by the Bordeaux brokers in 1855 for a fair in Paris and classified by them. These are among the top Bordeaux wines, starting with the five first growths, including chateaus Latour, Margaux, and Lafite-Rothschild.

The McLaughlins do enjoy other wines, but usually in restaurants. Recently they have been drinking more California wines. "For years, we were opposed to California reds," Mr. McLaughlin said. "But in recent years, we think they've improved. And we're actually drinking some of them, mostly in restaurants."

### Gaffelière, Giscours, Gloria, Greysac . . .

In spite of their unalterable loyalty to Bordeaux, the McLaughlins sense a loss of quality in recent years. "Bordeaux in the past few years has gone downhill," Mr. McLaughlin said. "We used to enjoy so many chateaus that had full bodies, almost a syrupy thickness. We rarely find them anymore. We're also surprised by the lack of complexity. We think Bordeaux is made now for quicker drinking. It simply isn't as good as it used to be."

That does not mean for an instant that the McLaughlins plan to change their wine-drinking patterns. "We tried to stretch the thing so as to include Burgundy," he said, "but we never got too far. Bordeaux really takes you in and holds you."

The couple have visited Bordeaux several times and have even met a few chateau owners, but outside of a few tastings run by Bob and Ben, they have had little to do with the so-called wine establishment in this country.

### Hanteillan, Haut-Batailley, Haut-Brion, Issan . . .

Despite their considerable experience with fine wines, the McLaughlins are not collectors. "We have a rack that holds about three hundred bottles," he said, "but we live on a flood plain and a couple of times the water has come in and covered the bottom two-thirds of the rack. There we were with two hundred bottles with no labels." Now their wine collection amounts to about one hundred bottles, all of them, it is hoped, above the high-water mark.

Mr. McLaughlin was concerned lest revelations about his modest obsession give the wrong impression to friends and neighbors. "I have a business here and my wife teaches school," he said. "We wouldn't want people to think we were, you know, alkies or something." He was assured that a couple of bottles of wine a week would shock none but the most hard-shelled teetotaler; that in fact, by almost any standard, he and his wife henceforth would rate as authentic connoisseurs.

"Oh, I don't know about that," he said, "but we do enjoy our little hobby. Which is just as well. There are more than six thousand chateaus in the Bordeaux wine region." The McLaughlins have a long way to go.

JUNE, 1991

## FOR ONE OREGONIAN, A LICENSE PLATE IS A CRUSADE

When Mike Higgins sold his wineshop in Oakland, California, in 1976, he retired to Oregon, an easygoing, sparsely populated state where personal freedom is still highly valued and everyone works hard not to step on anyone else's toes.

Or so he thought.

"I love it here," he said of the little town of Jacksonville, where he chose to settle, "although I especially miss my old winetasting friends and the customers at my shop, Vino!"

To keep alive the memories of his wine merchant days, Mike Higgins decided to apply for a license plate for his car that would read VINO.

Back came a letter from the Oregon Division of Motor Vehicles, turning him down. Why? Because in the Oregon agency's opinion, "vino" means "wine," and "wine" is a "drug-related word."

Mr. Higgins, who had seen many wine-related license plates in California, appealed, testifying at a State Transportation Department hearing in Medford. He asked permission to use INVINO, VINO, or WINE on his license plate.

Administrative Law Judge Teresa Hogan turned him down. VINO, the judge ruled, "would cause public offense or embarrassment to the state." Further, Mr. Higgins's driving around with WINE or VINO on his license plates "could convey the message that the state condones the use of alcoholic beverages by drivers" and "could well offend those citizens who have been victimized by intoxicated drivers."

"Wine" is banned from display on an Oregon license plate under a motor vehicles regulation defining drug-related words as those "which by denotation or connotation refer to any intoxicating liquor or controlled substance or their use." (Other no-nos under Oregon rules: "words of a vulgar nature, sex-related words, excretory words, words related to intimate body parts and ethnic words.")

"To include 'wine' as part of a definition of drug-related words while permitting 'smoke' and 'cigar' reminds me of Humpty Dumpty," Mr. Higgins said. "Humpty Dumpty said, 'When I use a word, it means just what I choose it to mean—neither more nor less.'"

Poking around in the state's files, Mr. Higgins discovered that a Champagne enthusiast was denied the use of "bubbly" and an orthopedic surgeon was denied "joint," which might have referred to his work or to a way of relaxing from his work.

Until 1993, Mr. Higgins learned, the Oregon motor vehicles agency also banned words of a religious nature. That November, an Oregon court held that "the regulations prohibiting issuance of a custom license plate with the word 'pray' are contrary to the free speech clause of the First Amendment," and ordered the plate issued. The state chose not to appeal.

Mr. Higgins found the Motor Vehicles Division's ruling in his case ironic in light of Oregon's efforts in other contexts to promote its important wine industry. He mentioned the Oregon Winegrowers Association, the Wine Advisory Board, the State Wine Cellar, and a state-published brochure, "Discover Oregon Wineries."

The Oregon Transportation Department, Mr. Higgins noted in a conversation, "even erects directional signs along state highways, which display those offensive, embarrassing words 'wine' and 'winery' for all to see."

He observed that wine has an important and honored place in the history, literature, art, religion, and culture of Western civilization, and in the economy of Oregon. The motor vehicles agency, he added, "should remove wine from this hurtful and unnecessary association with illegal drugs, vulgar words, and words that ridicule an ethnic class."

Undeterred by the state's rulings, Mr. Higgins plans to continue his crusade for the license plate. "I am placing my faith in the First Amendment and the judicial process," he said. "It will involve considerable expense to me," he added, "but 'wine' and 'vino' will appear on Oregon license plates."

In rejecting Mr. Higgins's contention that the motor vehicles regulations are an unconstitutional infringement on the right of free speech, Judge Hogan cited two California rulings that custom license plates were a "nonpublic forum," subjected to "a degree of regulation greater than that permissible with respect to pure speech."

Okay, so what kind of plates does California issue to its wine lovers? Well, there are plenty of "vino" plates: VINO 1, VINO 2, and VINO MIO, to name a few. Under "wine," there are WINE JUG, WINE OHH, WINE 4U, WINE 4ME, and WINE 4US. There is WINE 2GO and there is WINE SAP. Others are WINE WMN, WINE WIZ, and WINE SNB, which surely refers to that despicable fellow, the wine snob. There is even a WINE 911.

A white wine fan has CHARDNY, and what appears to be a trendy gang displays MERLOT 1, 2, 3, and 6.

Bourbon has its license plate partisans, and on a somewhat earthier plane, so does beer. Consider BEER GUT, BEER HOG, BEER HAG, BEER KEG, BEER LVR, and BEER ME.

Zinfandel, the grape that California has adopted as its own, turns up frequently, often in a play on its resemblance in sound to "sin": ZINNER and ZINFUL, for example.

There is a bit of zinfandel in Oregon, but not much. Which is just fine with Mr. Higgins because, in what must be disdain for most things Californian, the Oregon Motor Vehicles Division has allowed him a plate that reads ZIN.

MAY, 1997

## THE SPIRIT OF GIVING

This is not exactly a Christmas story; it's not really a wine story, either, although wine is central to it. It's a story about living and dying and the occasionally surprising nobility of the human spirit. No, it's not exactly a Christmas story, but for me it evokes the spirit of Christmas more profoundly than a hundred recorded carols or a forest of lighted trees.

The story begins in Norway in 1908. That was the year that Peder Knutsen, then nineteen, emigrated to Canada to homestead on the vast western plains. Sometime later, he wandered down into North Dakota, settling in Kindred, a farming community of some six hundred people that is twenty miles or so south of Fargo. He owned a bit of land, about two acres, worked as a farm laborer for his neighbors, and made occasional trips back to Norway to visit a sister. He never married, and toward the end of his life lived alone and apparently content in his small house, mostly on a $78 monthly pension from World War I. He died on November 11, 1974.

Several months later, on February 7, 1975, a small item appeared in the *International Herald Tribune* in Paris. It said that Peder Knutsen, of Kindred, North Dakota, had died recently and left a will donating $30,000 to a home for the elderly in the town of his birth, Gol, in the Hallingdal Valley, about 140 miles east of Bergen. But it was no ordinary bequest. Mr. Knutsen specified that the income from his bequest must be used to buy wine for the old people in the home.

On February 5, two days before the newspaper item appeared, the municipal council in Gol had agreed to accept Mr. Knutsen's gift ". . . even if the donation was odd and put this alcopolitical problem on our neck," as the acting mayor at the time, Ola Storia, put it.

Mr. Storia and his fellow councilmen estimated that Mr. Knutsen's bequest would provide about one thousand bottles of wine a year, or about three bottles a day to be shared among the twenty-three residents then living at the home. Assuming that there are always a couple of nondrinkers, even in a group of Scandinavians, the bequest barely came to a glass a day for each resident.

But a glass of wine is just what many physicians prescribe for their elderly patients, and even one glass of wine, shared with others at the evening meal, might serve to brighten some long arctic nights.

Shortly after the story about Mr. Knutsen's will first appeared, a neighbor—in fact, his landlord—spoke about him. "He was a wonderful old man," said Delice Ebsen. "He never went to church, but he was a religious man. His Bible sustained him."

The Ebsens had bought Mr. Knutsen's two acres from him around 1970 for $3,000. They let him live on, free, in the house he had built. "He used to make rhubarb wine," Mrs. Ebsen recalled, "and he enjoyed a can of beer now and then." So far as she knew, he had had little experience with grape wine at all.

"We talked about what he would do with his money," Mrs. Ebsen said, "and he liked the idea of leaving it to a home. But we had no idea he meant to leave it to a home in Norway, and to tell the truth, we had no idea how much money he had. There was the $3,000 he got from us, and he must have saved some money after selling his land in Canada."

The story of Peder Knutsen might never have come to light had it not been passed along by another elderly gentleman, George C. Sumner, an American, now deceased, who lived in Paris at the time. A Yankee of distinguished lineage, sent forth by Harvard in the misty years before the Great War, Mr. Sumner carved out an impressive career as a wine exporter and broker in the decades following World War II.

Appended to the now crumbling clipping about Mr. Knutsen are a few lines by Mr. Sumner: "It is not hard to imagine the drab dullness of life for those whose working days are over and who are confined to the routine of a home and the same old conversation—if any—at meals. A few glasses of wine and the fellowship they incite can change the atmosphere of the dining room and make dinner something to look forward to. I believe there are some, perhaps many, who would like to leave such an endowment for those who have had their day but still have some time."

It was Mr. Sumner's idea that a fund might be raised to provide a glass of wine for other old people living out their years in days of dull routine. How well he succeeded, I have no idea. I hope he did. But even if he didn't, it's still pleasant to think, particularly at this time of the year, of Peder Knutsen and to raise a glass to him and the old people of Gol in Norway.

Surely most of the old folks who first enjoyed his largesse have joined him in another life. He lived alone with his Bible in North Dakota, but thanks to him, succeeding generations of the elderly in a little town five thousand miles away have, from time to time, enjoyed a moment of warmth and friendship over a glass of wine.

DECEMBER, 1984

# CHAPTER 9

# RAISE YOUR GLASSES HIGH!
# ADVICE ON WHAT TO DRINK
# WITH WHAT (AND WHAT
# TO AVOID ALTOGETHER)
# AND SPECIAL THINGS
# TO WATCH FOR

Wine tasting can be an academic subject, with lots of note-scribbling and frowning; it can also be light-hearted and relaxed. After all, we don't go to school to learn how to eat bread. (Well, I never did.) There are ways to get to know and to evaluate wines without enrolling in a local university.

The location is important. Lloyd Flatt, a New Orleans businessman and once an avid wine collector, used to stage grandiose tastings at his home in the French Quarter to which people often traveled from France, Germany, or Great Britain, so rare—and plentiful—were the wines he offered. Actually, my all-time favorite tasting was one I conducted in the African bush—in Botswana, to be precise. Ah, the aroma of cabernet and elephants.

We take a look at garage wines, a relatively new phenomenon in the wine world, newer even than tastings on safari. They are Bordeaux wines made in minute quantities in small wineries sometimes referred to as garages. Several actually are garages. The winemakers are mostly young mavericks who also work in more conventional wineries. They want to break away from the constraints imposed by traditional Bordeaux methods. Garage wines are dark, almost opaque, intensely fruity, high in alcohol, and very expensive. Then we move on to pinot noir, both from this country and from Burgundy.

Also in this chapter is a rather choleric view of California winemaking I expressed almost twenty years ago. It seems a bit smug and overdone now, but it made some valid points back then. It must have; it got everyone in California sore as hell.

## TO ENJOY A GLASS,
## LESSONS ARE NOT NEEDED

Drinking wine is fun. Tasting wine is work.

And yet for some reason, Americans love to talk about wine tasting: "The meeting will be followed by a wine tasting." "After the candidate speaks, there will be a wine tasting in the library." "Join us for a wine and cheese tasting."

The wine broker in Bordeaux, who prowls the cold damp cellars trying to guess what the raw young wines will be like in another ten years, is a true taster. The importer confronted with, and probably confounded by, an array of seemingly identical bottles is a taster. So is the judge at a wine competition, facing one hundred cabernets. So is the winery owner, making the final crucial blend of the vintage. They all know what they're looking for.

These people are professionals who spend years developing their skills. They taste often and they taste seriously, because many thousands of dollars ride on their decisions.

Wine-tasting classes abound, particularly in the big cities, and they are serious enough—but mostly about meeting other tasters. As for all those other so-called tastings, they are mostly, to use the diplomat's term, drinkings. In other words, parties at which wine is served.

How can you tell a real tasting from a fashionable one? Simple: At a professional tasting, nobody—at least nobody with any sense—drinks anything. Since the nose is a far better judge of wine than the mouth, the pros spend as much time smelling wine as they do tasting it. And what they do taste, they spit out. Spitting is no fun. It's awkward, it's messy, and it's frustrating. After all, one's instinct is to swallow, particularly if the wine is any good.

Characterizing a wine and cheese party as a tasting is harmless enough, except for the fact that it conveys the idea of something educational, of a learning process. It implies that wine is something for the initiated; that to drink and enjoy it, you have to go through—yes—tastings.

I'd like to have a bottle of good Burgundy for every time someone has said to me, "I like wine, but I don't know anything about it." Why is it that no one ever says, "I like rye bread, but I don't know anything about it"? Why are so many so intimidated? Why do people think they have to take a course to be able to enjoy a glass of wine?

The industry bears much of the responsibility. How many kinds of rye bread confront the consumer every day? A dozen, if that? How many zinfandels? A couple of hundred. How many chardonnays? Probably close to a thousand. For all their vaunted marketing skills, the people who sell wine can't seem to do much about the resulting confusion.

The idea behind "tastings" appears to be that you will learn to distinguish among various wines. But who can make a choice among a thousand chardonnays? The wine publications are not much help. "We tasted 350 chardonnays for you," they trumpet, "and picked the 30 best." Thanks a lot.

A colleague came to me twenty years or so ago and asked: "Is Château X supposed to be a good wine?" He had had a bottle the night before and liked it. I don't remember the chateau name, but it was a good wine and at a good price. Years later, he said it was the only wine he ever drank. "I still like it," he said. "Why change?"

Why indeed?

If everyone had half a dozen wines he or she knew and enjoyed, much of the insecurity that accompanies wine drinking and wine buying would disappear, along with the lucrative "wine education" business that profits from it. Rare and expensive wines would continue to attract the relatively small market of collectors and connoisseurs, but a vast new market would open up for all the people who would like to drink wine but are put off by the confusion and elitism that plague the wine trade.

Almost lost in the approach to wine as something exotic is the idea of wine as food. The typical wine advertisement shows a handsome couple relaxing by the fire, sipping wine. To me, there is no greater waste of a good red wine than drinking it like a cocktail, either at home or at a bar.

In traditional wine-drinking countries, wine is almost never served as an aperitif or an after-dinner drink unless it is fortified. If you're just having a drink, have a martini or a beer or a Sherry. Or—the one exception—an oaky, sweetish California chardonnay. But save the red wine, or the dry white wine, for dinner.

Wine is meant to go with food. When Baron Eric de Rothschild took over Château Lafite-Rothschild in 1977, he invited a few people to come to the chateau for a long weekend to sample every vintage produced since the Rothschilds bought Lafite in the nineteenth century. As many of the wines as possible were tasted at the table, with meals.

Once or twice a year, Bipin Desai, a California collector, puts on extensive tastings of famous wines—every vintage of Ridge Vineyards' Monte Bello cabernet, for example—for other collectors. Mr. Desai's tastings, usually spaced over three or four days, are always done in restaurant settings. Like Baron Rothschild's weekend, they are not really tastings but a succession of resplendent meals.

Good wine and good food. Now that's what I would call fun.

MAY, 2000

## A TASTING OF LAFITES

Some years ago Baron Elie de Rothschild, then managing partner of Château Lafite-Rothschild, was given a very old Lafite to taste. "Well," he said, after a pause, "it's wine."

This seemed like a rather banal response at the time, but as anyone who has tasted many old wines knows, it was rather a compliment. It was also a description that proved quite useful in New Orleans in 1988 at what must have been the most spectacular tasting of Lafite ever assembled. The assembler was Lloyd Flatt, an aerospace consultant and an internationally known wine collector.

In 1977, when Eric de Rothschild took over the direction of Lafite from his uncle, Elie, he celebrated the occasion with a three-day tasting, at the chateau, of all the significant vintages since Baron James de Rothschild bought the property in 1868. It was a memorable event. A lunch consisted of lamb chops and a little cheese, accompanied by Lafite 1959, 1934, and

1873. With dinner—guinea hen—the wines were Lafite 1953, 1945, 1918, 1893, 1869, and 1921.

Mr. Flatt, unconstrained by dates of ownership, went back somewhat further in his three-day affair. His oldest wine was 1806, and the tasting included seven other wines from the pre-Rothschild era: 1832, 1844, 1846, 1848, 1858, 1864, and 1865. In all, he served 127 wines.

Hardy Rosenstock, a German collector, brought with him one of his "Thomas Jefferson" Lafites, dated 1784. Mr. Rosenstock created a sensation several years ago when he disclosed that he had acquired some eighteenth-century wines, including Château Lafite and Château d'Yquem, that had once been in the cellar of Thomas Jefferson. The bottles themselves, apparently very old and handblown, all bear the engraved monogram "TJ." Mr. Rosenstock has steadfastly refused to discuss the provenance of these wines, except to say that he bought them from a "Parisian collector," and their authenticity has been questioned.

In any event, the wine in the Jefferson bottle was vinegar. Which is why Elie de Rothschild's comment, "It's wine," was not as patronizing as it might seem. That wines more than a century and a half old are still wine is a remarkable phenomenon.

The oldest wines were tasted near the end of the first day. The group of four included 1806, 1832, 1844, and 1846. The 1832 and the 1844 were in poor condition; the 1846 was thin, pale, and weak, but it was wine, an authentic antique. The 1806 was still drinkable when the bottle was opened but faded rapidly in contact with air. In its brief exposure, it conveyed a bouquet of peach preserves. Mr. Flatt had bought the bottle from Christie's, the London auction house, in 1979, and had had it in his cellar since. Christie's had purchased it from the Restaurant Darrozes in Villeneuve-de-Marsan, in France.

The group consisting of 1848, 1858, 1864, and 1865 was tasted just before lunch on the second day. The 1858 was clearly the best of the group. In fact, the 1858 turned out to be a star of the entire tasting. Bright and clear, with a deep ruby color and a rich full taste, it could have been a wine from the 1960s. The 1848, too, was still a very drinkable wine with a full, sweet bouquet. The 1864, still considered one of the greatest Lafite vintages of the mid-nineteenth century, did not fare so well in this instance. Initially it had an attractive bouquet, but it declined rapidly. The 1865 was well past its prime.

Mr. Flatt's tasting was unusual in many ways, among them its depth in long-forgotten and mostly unobtainable vintages. He had, for example, every vintage from 1910 through 1929 with the exception of 1915. He had, as well, every vintage from 1937 through 1985, including the war years of 1940, 1941, 1942, 1943, and 1944, when German officers were billeted at the chateau, and the two poor years of the 1960s, 1963 and 1965, which most connoisseurs ignore. Of the 122 vintages beginning with 1864, only 13 were missing.

Also included in the tasting were eight vintages of Carruades de Lafite, the chateau's second wine, now known as Moulin de Carruades. The Carruades is a plateau that runs from Lafite into the adjoining Château Mouton-Rothschild and was once owned by Château Mouton d'Armailhac, now known as Mouton-Baronne-Philippe. While the Carruades is one of the best parts of the Lafite vineyard, the name was used for wine from younger vines. The practice was discontinued in 1967 but was revived in 1974 with the Moulin de Carruades, a name that was once used for a third wine made at the chateau.

Mr. Flatt's Carruades included 1902, 1922, 1923, 1934, 1937, 1961, 1962, 1964, and 1967. In some years, like 1922, the Carruades held up better than the first wine. For the most part, the Carruades proved to be fine wines, but coarser than the first wines.

Among the outstanding wines of the Lafite tasting, along with the 1858, were 1870, 1875, 1880, 1886, 1895, 1904, 1916, 1929, 1934, 1945, 1959, 1961, and 1982. Standouts were the 1858, 1875, 1929, and 1961.

Many of the older wines had been recorked at one time or another, some of them twice. Often, when an old wine is recorked it is "topped up" with wine from the same vintage, if possible, or with a wine from the same epoch. At one time, according to Michael Broadbent, the director of wine sales at Christie's, the 1865 vintage was a favorite for topping up old wines at the chateau. Some experts oppose recorking, saying it exposes fragile old wines to too much air.

The directors of Château Lafite-Rothschild are strong proponents of recorking. Periodically they send a team from the chateau to this country and anywhere else in the world where there are Lafite collectors. At no charge, they top up and recork old vintages of Lafite.

Each day of the tasting, for lunch, Mr. Flatt served yet another Lafite, but only in large bottles, double magnums or jeroboams. The lunch wines were 1964, 1960, and 1958. For the closing dinner on the third night, the wine was 1967 Lafite in magnums. The only Lafite wine not served over the three days was Lafite Blanc, which is made for consumption at the chateau and by the owning partners and is never commercialized.

There was one ringer: For anyone fatigued by so much red wine, Mr. Flatt also provided a white wine for lunch on the third day. It was a Burgundy at that, 1971 Bâtard-Montrachet in magnums from the Domaine Delagrange-Bachelet.

OCTOBER, 1988

## STALKING ZINFANDEL IN AFRICA

Does zinfandel go well with kudu? Or leopard? Elephant? And baboon, what about baboon?

Here in the far southwest corner of Botswana, in southern Africa, zinfandel goes with any of a hundred different wild creatures.

Not, to be sure, while eating these exotic birds and beasts, because in this game reserve, no one is allowed to harm them, much less eat them. Let's just say, they don't seem to object to anyone drinking wine in their company.

This fact became vividly clear at a "wine weekend" held at Mashatu in 1993, during which this correspondent conducted, if that's the word, a tasting of American wines deep in the African bush.

Mashatu, named after a tree native to this part of Africa, is the largest privately owned game reserve in southern Africa, comprising ninety thousand acres, twice the size of Brooklyn. Located about three hundred miles north of Johannesburg, it is bordered on the northeast by Zimbabwe and on the south and southwest by South Africa.

Although privately owned (by Rattray Reserves, a South African company), Mashatu is part of the 17 percent of Botswana's land devoted to wildlife preserves. Landlocked Botswana is supported by a substantial income derived from diamonds. Larger than France but with a smaller population than Dallas, Botswana is sometimes called the Switzerland of Africa because of its political stability and relatively strong economy.

Tourism is a growing business in Botswana. Travelers from around the world come here, though still in small numbers, to catch a glimpse of what many ecologists fear is a disappearing Africa. Visits of any length can be arranged, but at a remote reserve like Mashatu, the schedule is attuned to long weekends. Guests, almost all of them from South Africa, arrive on Thursday or Friday and leave on Sunday. Some drive to the South Africa–Botswana border and make the trip into the reserve in Land Rovers; most fly into the Mashatu airstrip in private planes or on charter flights.

From five to nine A.M. and again from four to eight or nine P.M., guests spend their time crashing through the bush in four-wheel-drive vehicles, tracking leopards, giraffes, cheetahs, impalas, kudus, baboons, and the herds of elephants, which trudge tirelessly through the park in search of food.

Animals and visitors tend to lie low in the midday heat, making it a good time for the pool—or for a wine tasting. On the first full day, the group went through a batch of interesting South African wines.

The next day, a steady breeze prompted the rangers to suggest taking the American wines outdoors. Carrying them gently in ice buckets over the rough, often nonexistent trails through the bush, we did, to an idyllic spot on the bank of the dry Limpopo River, a victim of southern Africa's decade-long drought. The Limpopo forms the border between Botswana and South Africa.

Onerous import duties combined with the now-lifted sanctions had effectively kept all but a minute quantity of foreign wine out of South Africa. So the twenty-five or thirty guests, all of them South Africans, were as eager to try a California chardonnay as they had been to see a kori bustard or lesser kudu earlier in the day.

What wines to show? Good wines, widely available in the United States at reasonable prices:

EDNA VALLEY VINEYARDS, CHARDONNAY, 1991.

BONNY DOON VINEYARD, VIN GRIS DE CIGARE, 1991.

SANFORD WINERY, PINOT NOIR, 1990.

RIDGE, ZINFANDEL, SONOMA, 1991.

FETZER VINEYARDS, BARREL SELECT CABERNET SAUVIGNON, 1989.

ROBERT MONDAVI VINEYARDS, RESERVE CABERNET SAUVIGNON, 1989.

Except for the Mondavi cabernet, which sold for more than $20, all the wines sold for about $15 or less at the time and were nationally distributed in the United States.

Burgundy-style wines are much in vogue in South Africa. Both new winemakers and older wineries are working to improve the quality of their pinot noirs and chardonnays. The Sanford Winery and Edna Valley wines were judged to be the equal of anything currently produced in South Africa, but even more impressive to the professionals in the group was the fact that thousands of cases of both wines were produced annually.

Most in the group felt the Bonny Doon Vin Gris, a rosé, was excellent but not much different from some of the better rosés produced in South Africa. The Mondavi reserve cabernet, though still closed-in, was greeted with something akin to awe, not so much for its intensity as for its elegance and breed, qualities sought after but not always achieved to such a degree in wines of the western Cape.

The popular favorites of the day were clearly the Fetzer cabernet and the Ridge zinfandel. Cabernet is not uncommon in South Africa but tends to be austere. The plummy richness and sheer accessibility of the Fetzer wine came as a revelation to the tasters. Even more astonishing was the price, usually under $10.

The zinfandel got much the same reception. A small amount of zinfandel is produced in South Africa, but it's still considered rather exotic. The tasters were intrigued by the fact that a wine could be so intense and so drinkable at the same time.

The late Peter Devereux, a Johannesburg writer and wine expert who presided over the wine weekend, told his assembled countrymen that the tasting, even though limited to half a dozen wines, indicated to him that South African winemakers were fifteen years behind their Californian counterparts. He spoke of South Africa's long isolation and need to catch up in order to share in the world market. "When South Africa can make wines like these consistently and at similar prices, we can be said to have arrived," he said.

Among other things, the tasting in the bush showed that good California wines can stand up under some unusual conditions. Before the jaunt into the bush, Mr. Devereux banned smoking and suggested that no one wear any strong scent.

He forgot to tell the elephants. Fortunately the breeze held up for most of the afternoon.

JANUARY, 1994

# A TWENTY-THOUSAND-BOTTLE WEEKEND

The next time you are trapped washing wineglasses after the guests leave, cheer up. After each big tasting at the New York Wine Experience in October 1999, members of the staff at the Marriott Marquis Hotel had to wash and dry some twenty-seven thousand of them.

Everything at the Wine Experience, a three-day extravaganza for wine buffs, was outsize, including the number of guests. About eleven hundred signed up for the whole long weekend, and an additional fifteen hundred turned up on either Thursday or Friday evening to sample some of the 350 wines that were being poured by representatives of 250 wineries worldwide. At several of the sit-down tastings, a thousand participants consumed a total of more than ninety cases of wine. Twenty sommeliers worked behind the scenes, opening, decanting, and pouring more than twenty thousand bottles of wine over three days.

The New York Wine Experience is presented every two years by M. Shanken Communications, the publisher of *Wine Spectator* and other magazines. In alternate years, the California Wine Experience, mostly featuring that state's wines, has been held in San Francisco. The Wine Experience—this was the nineteenth—is the largest regularly held wine event in this country, but it is hardly a recent phenomenon, nor is it the largest worldwide.

That distinction probably goes to the Fête des Vignerons in Switzerland, which dates back to the fourteenth century. It is held every twenty-two years. A theatrical spectacle as well as a wine fair, it attracted more than a million visitors to the town of Vevey over a three-week period in July and August 1998.

The New York event also offered a banquet on Saturday, featuring the singer Gladys Knight, but the main entertainment was the promotion of wines. Most of those who attended the Wine Experience were relatively sophisticated wine drinkers; not too many neophytes are likely to part with $950 for the weekend. And producers eagerly showed them—and got them to drink—their wines.

That the participants were fairly knowledgeable was evident from the exhibitors they sought out at the Grand Tastings. A crowd at the Château Mouton-Rothschild booth was no surprise, especially with Baroness Philippine de Rothschild pouring, but almost as many eager fans jostled for a chance to sample Ridge Vineyards' Monte Bello cabernet sauvignon from the Santa Cruz Mountains, the *premier cru* Puligny-Montrachet of the Domaine Leflaive in Burgundy, and Unico, the legendary wine of Vega Sicilia in Spain.

Though guests who attended only the two Grand Tastings paid $165 each, and showed determination by waiting in long lines, they were able to taste wines worth considerably more than they paid. Besides Mouton-Rothschild, Bordeaux was represented by Châteaux d'Yquem, Margaux, Lynch-Bages, Rauzan-Ségla, and Vieux-Château-Certan, among many others. Rare Italian wines like Angelo Gaja's Alteni di Brassica sauvignon blanc and Montevertine's Pergole Torte, from Tuscany, were available.

In the early days of the Wine Experience, many exhibitors were content to show their less expensive wines. Now the Wine Experience audience demands the best, and in most cases, they got it. Instead of pouring its regular brut Champagne, Veuve Clicquot poured its top-of-the-list Grande Dame. Instead of a tawny or some lesser Port, the Quinta do Noval poured its vintage 1995. Each exhibitor paid $1,200 to participate and was required to bring six cases (seventy-two bottles) of a single wine.

One enticing tasting was on Saturday morning, when guests were offered samples of wines from some of California's smallest and most sought-after producers, including Patz & Hall, Testarossa Vineyard, Flowers Vineyards and Winery, David Arthur Vineyards, and Howell Mountain Vineyards. Virtually unknown to the public, these "cult" wines are the ultimate goal of wealthy collectors. As a group, the cult wines are beautifully crafted and long-lived. But their high prices reflect their scarcity as much as their quality. Some of these wineries are so small that they had to part with almost half a year's production to cover the Wine Experience's needs.

A tasting on Saturday afternoon featured wine from five of the so-called super seconds, Bordeaux chateaus that match or better the wines of some of the famous first-growth chateaus like Margaux and Latour. Featured were the quite expensive wines of Châteaux Palmer, Léoville-Barton, Pichon-Baron, Pichon-Lalande, and Cos-d'Estournel.

The Wine Experience is a nonprofit event. Income goes to a scholarship fund to assist people studying for careers in the wine industry. Beginning in 2000, the California Wine Experience will move from city to city nationwide. Its first stop will be New York. Bringing the California event to the East will give New York wine lovers a chance to experience some 150 California wines not normally poured at the New York Experience.

Among the Wine Experience's biggest fans are the exhibitors. The show gives them a chance to meet many of their competitors and taste their wines, and to visit New York restaurants and perhaps drum up some trade.

The same is true of the sommeliers, most of whom work for well-known restaurants. Saskia Schuit, who has served as a sommelier at every Wine Experience for twelve years, flies in from Amsterdam, where she is an importer and brings in many California and Washington State wines. "This is where I find out what's going on in my trade," she said.

Is the Wine Experience worth $950? The heat and the crowds can be oppressive in the Grand Tastings, and the long corridors of the Marriott Marquis bear little resemblance to the Robert Mondavi Winery or Château Lynch-Bages, but there is probably no better way to sample so many of the world's best wines in one location.

The three-day program was sold out four months ahead, and the evening tastings a month ahead. Once $950 would buy a cellarful of wine. Today it barely buys a case of good Bordeaux. Which probably makes the Wine Experience a pretty good deal.

OCTOBER, 1999

## $1,000 WINES YOU NEVER HEARD OF

The vines, bereft now of their grapes, are turning red, gold, and russet, and the brisk autumn air is spiced with the bouquet of new wine. "A Medieval City," say the placards on the town hall, and indeed in St.-Émilion, a sleepy wine center thirty miles east of Bordeaux, little seems to have changed in five centuries.

Nevertheless, there is a revolution going on here, a wine revolution that has shaken the foundations of the most important region in France. And like all revolutions, it began with a small group of people who had the audacity to question the way things were always done. The group bears the unlikely name of *garagistes,* a label manufactured for people who make wine in garages. None of them actually do, but insofar as the name sets them apart as iconoclasts, it is accurate.

Simply stated, the *garagistes* make exceptionally good wines in exceptionally small quantities and sell them for exceptionally high prices. The wines are called *vins de garage,* or "garage wines," and the most prized are selling for more than Pétrus.

There are at least a dozen made here in St.-Émilion, but garage wines are now also cropping up in the Médoc, the most traditional of all the Bordeaux regions; in the Graves, south of the city of Bordeaux; and in Italy and Spain. Buyers, often on the Internet, are fighting to spend as much as $1,000 a bottle for wine that, in some instances, no one ever heard of before.

And now the whole wine world is caught up in a swirling controversy: Are these great modern wines made by people willing to break the mold, or are they overhyped trophies for the moneyed collector?

The original garage wine comes from Pomerol, the wine commune adjacent to St.-Émilion and the home of Pétrus, until recently the rarest and most expensive of all Bordeaux wines. It is called Le Pin, which means "The Pine Tree." Its entire vineyard is only about five acres, and it rarely produces more than six hundred cases of wine a year. The wine cellar is in a battered old farmhouse on the plateau of Pomerol, less than a mile from Pétrus.

Like many of these "new" winemakers, Le Pin is owned by a family with deep roots in the wine business: the Thienponts, Belgians who have owned chateaus in Bordeaux since the 1920s. Their best-known property is Vieux Château Certan, which adjoins Le Pin. When the Thienponts bought the tiny plot that became Le Pin in 1979, they planned merely to extend the Vieux Château Certan vineyards.

"When we realized that Le Pin soil was very special, we decided to keep it separate," Jacques Thienpont said. "So we created Le Pin." In 1985 the

Thienponts acquired another adjoining hectare (2.5 acres) from the local blacksmith.

In the early 1980s, word began to spread that an astonishingly good Pomerol was being made from a tiny vineyard just down the road from Pétrus, and with the 1981 vintage Le Pin began to take on cult status. When its wine began to surpass Pétrus in important tastings, it became the hottest property in Bordeaux—in France, some would say.

Other winemakers took notice. By the mid-'90s, almost a dozen new wines had appeared, all of them made in Le Pin style—rich, full-bodied, and packed with flavor—all made in extremely small quantities and most of them made here in St.-Émilion, next door to Pomerol.

Since a number of these new wines were produced, not in great chateaus but in modest digs similar to Le Pin, a French writer, Nicholas Baby, dubbed them *vins de garage* and the people who make them *garagistes.* Fiona Morrison Thienpont, who runs Le Pin with her husband, Jacques, bridles at the name.

"We are not *garagistes,*" she said. "We were making wine here a dozen years before anyone thought up the name."

I found hardly any other *garagiste* who was happy with the title. And in fact, there is nothing casual or make-do about these *microcuvées.* The formula is simple: small quantities of the best grapes; severe selection, which means cutting away excess bunches during the growing season and retaining only about half of the harvested grapes; new oak barrels every year; and hands-on care up to and including the bottling.

Among the best-known *garagistes* here in what is known as Bordeaux's Right Bank (of the Garonne River) are Valandraud, La Mondotte, La Gomerie, Barde-Haut, Pavie Macquin, Gracia, Grand Murailles, and Tertre Rôteboeuf, all of them in St.-Émilion. In the Médoc, the home of the greatest Bordeaux chateaus, a newcomer named Marojallia, from the commune of Margaux, is priced even above Château Margaux itself.

In the Ribera del Duero, in northern Spain, Pingus, by the Danish enologist Peter Sisseck, has created as much sensation as any St.-Émilion wine, selling in Spanish restaurants for $500 a bottle. In Italy Palazzi, from the Tenuta di Trinoro, and Lamborghini Campoleone have had the same impact.

In irreverent Australia, a garage wine made from shiraz—syrah in the Northern Hemisphere—is called Duck Muck.

All of these wines owe some of their popularity to the success of the so-called cult wines of California, themselves mostly a phenomenon of the late 1980s and early '90s. With wines from microscopically small properties like Harlan Estate, Bryant Family Vineyards, Screaming Eagle, Marcassin, and Araujo sold only through mailing lists and bringing four-figure prices when they are resold, it was probably only a matter of time before winemakers elsewhere caught on to what could be done in such affluent times.

Whether St.-Émilion garage wines or California cult wines, what sets them apart from the crowd is not so much their style or their quality, which after all can vary from vintage to vintage. It is, perversely, their unavailability that makes them so desirable.

Serious wine drinkers will pay a lot for the finest wines; collectors will pay that much and much more. The price for most of these wines at the cellar door, so to speak, is high but manageable. Unfortunately, the wines are all spoken for before the grapes are crushed. People have been known to offer to sell their place on a cult wine list.

For most of the world, the only places to go to buy these wines is the auction room or auctions on the Internet, where they are knocked down at staggering prices. At a Sotheby's auction in November 2000 in New York, four bottles of the 1982 Le Pin were offered at a starting price of $5,000 to $7,000; five bottles of the 1990 at $3,000 to $4,000; and three bottles of Harlan Estate 1995 at $900 to $1,200.

Will any of these wines ever be drunk? If the buyers have a chance to profit from reselling, they are not about to drink their good investment.

The Thienponts insist they had no idea that Le Pin would take off as it did, and their original prices were moderate by Bordeaux standards. In the early '80s it was possible to find a case of Le Pin for $450. Today at auction, which is the only place the wine ever comes to market, a single bottle can sell for $600 or more.

The next of these wines to make an international reputation was Valandraud, which Jean-Luc Thunevin makes in a beautiful little winery—it may once have really been a garage—hidden away in a narrow cobbled street in the center of old St.-Émilion. Valandraud first appeared in 1991. Tall, diffident, with a quiet sense of humor, Mr. Thunevin comes from an old Bordeaux wine family. It is his wife who makes the Margaux garage wine Marojallia.

La Mondotte, which became a garage wine in 1996, is made by Count

Stephan von Neipperg at one of his other properties, Château Canon-La-Gaffelière, also in St.-Émilion. Like many of these hot new wines, La Mondotte is not new at all. It appears in records dating back to the middle of the nineteenth century and may well have existed for a century before that. It had fallen into disrepair when the von Neippergs bought it in 1971.

"La Mondotte was listed in an 1868 directory of chateaus," the count said. "Originally, I thought of adding its production to Canon-La-Gaffelière. Only when the authorities turned me down did I decide to develop it separately. I'm very glad I did."

The von Neippergs, who also own Château Peyreau and Château Aiguilhe in St.-Émilion, trace their history as part of the Franconian nobility back to the twelfth century. "My family has been making wine for eight hundred years," Count von Neipperg said.

Château Barde-Haut is another "new" wine that has been around for a long time. In fact, its present owners, the Garcin-Cathiard family, have owned it only since the middle of September 2000. Its previous owner, having developed the property along *garagiste* lines, found himself in trouble with French tax inspectors, sold off his holdings, including Barde-Haut, and moved to Monaco. The Garcin-Cathiards, originally from Grenoble, also own Clos l'Église in Pomerol, Haut-Bergey in the Pessac-Léognan part of the Graves, and considerable vineyard acreage in Argentina. A brother, Daniel Cathiard, owns Château Smith-Haut-Lafitte, also in the Graves, which is becoming as well known for its beauty spa as for its wines. Barde-Haut is a small chateau but a particularly beautiful one.

Hélène Garcin Leveque, a daughter of the Garcin-Cathiards, is charged with running Barde-Haut. She and her husband, Patrice Leveque, who also comes from a chateau-owning family, are developing yet another garage wine on a tiny plot his family has in the Graves.

La Gomerie, also in St.-Émilion, emerged as a garage wine in the mid-'90s. Once the private vineyard of an abbey, La Gomerie is owned by Gérard and Dominique Bécot, who also own the adjoining Château Beau-Séjour Bécot. La Gomerie is said to be the Bécots' attempt to make one of the best of all St.-Émilion wines.

Not unexpectedly, in a hidebound region like Bordeaux, the garage wines have provoked considerable controversy. Partly it involves the people who make them, who are often seen as newcomers, insensitive to Bordeaux's tradition of making subtle, less forward, complex wines. Partly it involves the enologist Michel Rolland, a longtime advocate of big-bodied, flamboyant wines, who has been a driving force behind many of these new labels, and partly it is a reaction to the powerful American critic Robert M. Parker Jr., who has been effusive in his praise of the garage wines and of Mr. Rolland, seemingly at the expense of the Bordeaux establishment and the great names it reveres: Lafite, Mouton, Ausone, Margaux, and others.

Bordeaux traditionalists maintain that these new wines have no track record. Bordeaux styles, they say, have developed over two and a half centuries and have made Bordeaux wines the standard for the whole world of wine. Who knows, they ask, how well these new wines will develop, how long they will last? Of course, no one knows. At the same time, the wines are immensely attractive, which should be reason enough to welcome them.

Still, it is well to note that they are not always as stupendous as they are made out to be. In comparing the 1998 La Gomerie with the 1998 Beausejour-Becot, I found the older house's wine clearly superior in elegance and in harmony or balance. The same was true of the Barde-Haut when compared with Clos l'Église, and La Mondotte with Canon-La-Gaffelière. I thought Le Pin equal, perhaps even a bit superior to, its sibling next door, Vieux Château Certan. But I tasted them separately, which is not as good as having them side by side. Valandraud, which I also tasted separately, was the most harmonious and perhaps a bit more subdued. A lovely wine.

Wood is very prominent in the garage wines, in some cases recalling California zinfandels that seem to use oak for oak's sake. To me, the worst offender in this respect among the garage wines was the Spanish wine Pingus, which seems to have been made mostly for wine competitions: great for vanquishing less assertive wines, but too big, too overpowering for the dinner table, even with steak, which is the way I had it.

The success of these wines will undoubtedly inspire many more. I heard of at least a dozen startups while I was in France and Italy in 2000. Some will be poor imitations, but others will be good wines in this same modern style. They should come as no surprise. Bordeaux has been coming around slowly, like a great ship in the center of a harbor, for years.

At a recent tasting of 1985 and 1995 Bordeaux and California cabernets, it was easy to distinguish the 1985 Bordeaux from the 1985 cabernets. With the '95s, it was tough to tell which was which. The Bordeaux had become startlingly similar to the California wines.

It is a trend that has been going on for some time. The garage wines have only made it more striking. Does it represent some globalization of taste, a McDonald's-like homogenization? Maybe. It could also signal a new generation of wine drinkers in Europe who will join Americans in discovering new wines and new styles of wine.

Inevitably, competition will bring down the price of these wines. And when there are many more of them, they won't seem nearly so impressive as they now do. Then they can be compared with the more traditional styles for taste, rather than cost.

They will do well, but I don't think the great old wines of Bordeaux have much to worry about.

OCTOBER, 2000

## BURGUNDY MEETS ITS MATCH: AMERICAN PINOT NOIR

After a half century of stumbling efforts and too many premature pronouncements of success, it can finally be said with certainty that the pinot noir produced in the United States has arrived as a world-class wine. The best pinots of California and Oregon are now the worthy rivals of the great wines of the fabled Côte d'Or in the heart of Burgundy.

The point was driven home to me vividly in 1998 when I had the chance to compare 1995 pinot noirs from two great producers, one in Burgundy and the other in California. The French wines were a Corton Pougets and a Beaune Clos des Ursules, both produced by Louis Jadot. The American counterparts, from the Williams Selyem Winery in Healdsburg, California, were an Olivet Lane Vineyard and an Allen Vineyard pinot noir.

There were differences. The French wines were not yet all they could be, holding back a little, but they were still elegant and harmonious. The California wines were ready, rich and flavorful. More to the point, these American wines, from the Russian River Valley of California, were just as subtle and complex as the Burgundies. Great pinot noirs, every one.

True pinot noir, whether from Healdsburg, California, Volnay, France, or Dundee, Oregon, is light in body and delicate but intense at the same time. It's not uncommon to pour a glass of fine pinot noir that looks almost like a rosé, only to find that it's astonishingly powerful in both taste and bouquet.

Beyond a native interest in this country's produce, there are solid reasons to hail the tremendous advances of the last few years. For one, there are some wonderful buys in the American pinots, some terrific wines for less than $10. To find the same quality in a French pinot you might have to pay double. What's more, buying Burgundy can be a crapshoot. Quality varies, not just from year to year but also from bottle to bottle. The American wines are more consistent.

There will never be too much great pinot noir, but there will be more in years to come: The American success with the grape has prompted courageous winemakers the world over to try their hand at it. Australia and New Zealand have produced creditable Burgundy-style pinots. Although nothing particularly exciting has come out of South America, South Africa is producing several excellent pinots.

The route to good pinot noir in this country was long and arduous. In 1946 and 1947, the enologist André Tchelistcheff made what he considered to be top-quality Burgundian-style pinot at Beaulieu Vineyard in the Napa Valley in California. But he died in 1994 at ninety-two, acknowledging that he had never matched those wines again.

Fortunately for pinot noir lovers, other American winemakers took up the challenge. There were false starts, a lot of heartache, and some dreadful wines.

For years the accepted wisdom held—Mr. Tchelistcheff's first effort aside—that the United States would never produce great wine from the pinot noir grape, which is susceptible to the slightest of variations in weather and temperature. The California climate was too hot and too dry, and the soil was all wrong. Even the Burgundians, who have been working with it for centuries, have trouble growing pinot noir and more trouble turning it into wine.

What's more, no one else in the world had had much luck with the grape, either. The English writer Jancis Robinson calls it "that minx of a grape," and James Halliday, a well-known Australian writer and winemaker, has accused it of being "a temperamental and fickle mistress."

Even so, American winemakers have refused to give up. There were experiments at Hanzell Vineyards, in Sonoma County, in the late 1950s; Louis M. Martini and Heitz Wine Cellars, both in Napa County, made respectable pinots around the same time. Chalone Vineyard, hidden in the mountains east of Monterey, began making good pinot noirs in the early 1960s and, almost alone, continued to make them through the '70s and '80s. But those were the macho days of the California wine industry, when bigger invariably meant better. Unfortunately, that rule doesn't apply to pinot noir. The delicacy is lost when the alcohol content goes up.

There was some precedent for the "big" stuff, however. In the late nineteenth century, Englishmen who had cut their teeth on Port demanded big wines. To satisfy this craving, wily Burgundian merchants laced their light-colored, elegant wines with thick red wine from Algeria and the Midi. Soon a generation of British topers was smacking lips and telling sons, "Now, *that's* Burgundy."

After World War II it took American growers twenty years to get back on track. The numbers bear this out. There were 10,200 acres of pinot noir in California in 1976. They shrank to about 7,700 acres in 1986. By 1995 there were about 9,000 acres, and current estimates put California pinot noir at about 11,000 acres. (By comparison, California currently has about 40,000 acres of cabernet sauvignon and 75,000 acres of chardonnay.) Oregon, whose winemakers led the pinot noir revival in the 1980s, has another five thousand acres of pinot noir vines. The Oregon factor is crucial in the development of pinot noir.

Starting from nothing in the late 1960s and early '70s, a handful of pioneers, some of them disillusioned by the growing commercialization of California winemaking, moved north. In the red soil of the Willamette Valley in Oregon, just southwest of Portland, they created an artisanal wine industry that, for a time, threatened to eclipse California entirely in the production of fine pinot noir.

The original Oregon growers went north because they had become convinced that California was too hot and dry to make Burgundian-style pinot. The Willamette Valley, they discovered, was much more like Burgundy in climate and soil. Later, the Calfornians began to find local pockets of the kind of climate and soil conducive to good pinot noir, and they soon caught up to the Oregonians. The rivalry is friendly but serious. The outcome: good pinot noirs from both states.

It was in Oregon, not California, that the famed Burgundian house of Joseph Drouhin decided to make its first New World pinot noir in 1988.

Almost all of Oregon's pinot noir is grown in the Willamette Valley, which runs north and south between Oregon's Coast Range and the Cascade Mountains. Most of the wineries are clustered in the northern end of the valley, not far from the growing bustle of Portland. Dick Ponzi's winery and vineyards at Beaverton, a rural suburb of Portland where he started thirty years ago, are now all but swallowed up by urban sprawl, much the way Château Haut-Brion was overwhelmed by the spreading city of Bordeaux fifty years ago. But in spite of his semi-metropolitan setting, Mr. Ponzi consistently turns out some of the country's best wines.

California's pinot regions are more diverse. In the north they include the Anderson Valley in Mendocino County, the Russian River Valley in northern Sonoma County, the Carneros region at the southern tip of Sonoma and Napa counties, parts of Monterey County, and a large portion of the central and south-central coasts, stretching from Paso Robles south to Santa Barbara.

Every one of these regions can and does produce fine pinot noirs—most of the time. Even the most successful pinot noir producer will agree with the wine writer Bob Thompson, who once said of the grape, "Even where it prospers, it needs to be coaxed, wheedled, flattered, cajoled, cursed and (or) prayed over almost ounce by ounce through a series of crises that starts at the fermenters and lasts beyond bottling."

Then there is the problem of style. Having spent more than a few years at this, in 1998 I took part in a large-scale blind tasting of pinot noirs from France, California, and Oregon. Among the tasters were people who had made some of the wines. As a tasting, it was a great success; as an ego trip, a disaster.

Oregon wines were identified by everybody as French, French wines as Californian, and Californian as Oregonian. Does this mean that growing conditions in California are the same in the Santa Maria Valley as in Carneros? No. But weather variables, soil variables, and winemaking idiosyncrasies still strongly influence wine style.

It will be years before all these elements in a wine are sorted out. In the meantime, the wines get better and better. Currently the most praised source of good pinot noir is the central coast of California and particularly the wineries that buy all or most of their fruit from two famous vineyards in Santa Barbara County: Sanford & Benedict Vineyard and Bien Nacido Vineyard. Some thirty or forty wineries buy grapes from one or the other, or both.

Au Bon Climat, Babcock Vineyards, Foxen Vineyard, Cambria Winery, Fess Parker Winery, Byron Vineyard and Winery, Sanford Winery, Talley Vineyards, and Wild Horse Winery are a few of the better-known and most accomplished vineyards from the central and south-central coast areas. From the Monterey Peninsula and the north-central area, some of the more prominent producers include David Bruce Winery, the Calera Wine Company, Estancia Monterey Estate, and Morgan Winery.

My own preferences, at the moment anyway, tend toward vineyards farther north. From Carneros, Acacia and Saintsbury stand out. From the Russian River area, Williams Selyem, J. Rochioli Vineyards, Gary Farrell Wines, and Dehlinger Winery are all excellent.

J. Stonestreet, La Crema, Robert Mondavi Winery, and Rodney Strong Vineyards are other outstanding pinot noir producers. Based in Lake County, north of Napa, Jed Steele produces an array of pinot noirs with grapes from top-flight vineyards all over California. His 1995 pinot noirs include one from Carneros, one from the Anderson Valley, and three single-vineyard wines: from Bien Nacido in Santa Barbara, the DuPratt Vineyard in Mendocino, and the Durell Vineyard in Carneros.

Others to look for include Landmark Vineyards, Marimar Torres Estate, and Iron Horse Vineyards, all Sonoma wineries, and two pinot noirs from Kendall-Jackson Vineyards, a Vintner's Reserve and a more expensive Grand Reserve.

Among my favorite Oregon pinots are wines from Ponzi, Adelsheim, Benton Lane, Erath, Sokol Blosser, St. Innocent, and Domaine Drouhin, all from the northern end of the Willamette Valley.

One problem with good pinot noir is its relative scarcity. Some of the wineries listed here, like Williams Selyem, sell only by mail order, which means that because of the spate of new blue laws being enacted around the country, only a few people, almost all of them Californians, will ever taste the wines. I mention them because they are truly superb wines and, well, you may find one in a restaurant now and then.

Happily, there are some good pinots out there that have been made in reasonable quantities and that for the most part are being sold at prices of less than $20. Count among them Rodney Strong's Russian River; Firesteed, from Oregon; Turning Leaf, a real surprise from Gallo; Estancia; Fetzer Vineyards; and Napa Ridge. They are labels that should not be too hard to find. Napa Ridge, a division of Beringer Vineyards, has a remarkable record for bringing in good wines at affordable prices. How Napa Ridge's wine-maker, David Schlottman, is able to produce such a high-quality pinot noir year after year, even when good grapes are in short supply, and offer it for around $9 is a mystery. How does he do it? Well, actually, who cares? As long as he doesn't stop.

## VIVE LA DIFFÉRENCE!

Herewith, ten outstanding American pinot noirs. The first five, expensive and in short supply, are among the very best in the world. The others are good values.

PONZI VINEYARDS RESERVE—*Willamette Valley, Oregon.*

WILLIAMS SELYEM WINERY—*Russian River, J. Rochioli Vineyards, Sonoma County, California. Sold only to restaurants and to those on a mailing list. To add your name to the list, phone (707) 433-6425.*

STEELE WINES—*Bien Nacido Vineyard, Santa Barbara County, California.*

AU BON CLIMAT—*La Bauge Vineyard, Santa Barbara County, California.*

TALLEY VINEYARDS—*Rosemary's Vineyard, Arroyo Grande Valley, San Luis Obispo County, California.*

KENDALL-JACKSON VINEYARDS—*Vintner's Reserve, various vineyards in California.*

TURNING LEAF—*Reserve, Sonoma County, California.*

ESTANCIA MONTEREY ESTATE—*Pinnacles, Monterey, California.*

MERIDIAN VINEYARDS—*Santa Barbara, California.*

KING ESTATE—*Willamette Valley, Oregon.*

# A DISSENTER'S VIEW OF
# CALIFORNIA WINES

I used to be among those who were constantly trying to write something perceptive and flattering about California wines. "The California miracle . . . thirty years ago an industry in ruins . . . up from the ashes of Prohibition . . . new wineries every week . . . better and better . . . more and more"—not original, but heartfelt.

Not surprisingly, a couple of years in Europe, away from that charged atmosphere, effected a few changes. Actually it started fairly early in my tour. Lonely travelers would appear bearing California wines in their kits. They—the wines, not the travelers—seemed to have lost some of their charm, the reds in particular. Was it some kind of latent snobbism? Was it possible to become the Henry James of wine, eager to embrace the Old World and shuck off the New?

Then there were a couple of trips home and the concomitant access to plenty of California wine. Same results. I found myself asking, reluctantly, what all the fuss was about. The wines were good, some of them, and often delightful to drink. But the unremitting enthusiasm of their fans and promoters, the breathlessness, seemed excessive. The miracle seemed a bit more commonplace; it appeared that, having expressed our amazement that the dog could talk, it was time to determine what it might have to say.

Ask a professional marketing man about wine statistics and he will dazzle you with charts: huge increases in purchases, in per capita consumption; America becoming a wine-drinking nation. Then ask him about what America drinks. Wine dwindles into relative insignificance. Another batch of charts will show that while it is true that consumption has gone up, it is also true that the number of people who drink wine has not.

In America wine has its own newsletters, its own stars, its own clubs, even its own vocabulary. Where else in the world would someone sip a little wine and then, with a straight face, remark, "I understand they brought these grapes in at 24 Brix."

What is or are "Brix"? Sorry, you are not a member. Cultists flourish not so much on shared joys as on imagined enemies. American wine enthusiasts dearly love the image of the French expert, white-faced and with fists clenched because he has just identified a California cabernet as a Château Mouton-Rothschild 1945.

Indeed, the French wine world—the people who make it, the people who sell it, and to a lesser degree, the people who drink it—are intensely interested in the American wine scene. They love to visit our wineries and vineyards, they love to welcome American winemakers here, and given the problem of the language barrier, they are happy to receive touring American wine drinkers. They just do not consider American wines particularly good. They do believe that American wines will one day be very good— but not for a long time and probably not from the vineyards we now consider among the best. I am inclined to agree.

This is no blinding Archimedean revelation. There were, for me, those years at the Los Angeles County Fair, judging literally hundreds of California wines—judging them with California vintners—and in many categories finding it difficult to come up with something drinkable. And these were wines their producers thought worthy of honors.

Recently a reformed newspaperman turned California wine fan appeared in Paris. Here is one of the stories he told: It seems that there is a small winery "up in the valley" whose output is highly prized by the cognoscenti, so much so that the winemaker-proprietor can sell everything he makes with his mailing list. My friend was ecstatic because he had been able to take over a place on the list from a friend who preferred drugs.

This particular winery does produce good wines, but not all are worth getting in line for. The notion that one winery will turn out consistently superior wines, year in year out, the way Cartier does watches, is absurd. This is not wine drinking; it is a kind of bush-league elitism, like clothes with someone else's name on them.

Not that elitism is confined to the customers. An astute Frenchwoman with years in the wine trade said recently: "I adore the California chardonnays, but I don't know what to do with them. They certainly don't go with meals." She was half right. They are meant to go with meals, but many of them do not. Like overbred dogs, they have gone beyond their original purpose. They are too aggressive, too alcoholic. They are showoff wines made by vintners who seem to be saying, "I can outchardonnay any kid on this block."

We took one of those macho white wines from a famous North Coast winery to a dinner at a three-star restaurant in the French countryside. We matched it with a famous Burgundy. Both were chardonnays, from optimum vintages and in the same price range. At first the California wine was impressive and the French wine seemed weak and bland. Twenty minutes into the meal, however, the American wine was clumsy and overpowering while the charm and subtlety of the French wine was only beginning to emerge.

If I mentioned the name of the California wine, a host of its devotees—well, two or three—would leap forward growling. "Why, that wine took top honors in my last fifteen tastings!" an opthalmologist would cry. Exactly. There should be a special label warning that says: "This wine was designed for competition and is not to be used for family dining."

What is this frantic tasting business all about, anyway? Is it not possible to enjoy wine without having won the equivalent of a black belt?

The point is this: The drinking of wine in the United States, particularly American wine, is on the brink of becoming inbred and precious. Wine enthusiasts, including we writers, who should know better, ape the jargon of the trade and feel special when we exchange arcane trivia about grape crushers, red spiders, and who is opening next week's winery. A continuing hyperbole competition distorts wines, places, and people beyond recognition.

One day the rest of the country, bemused and probably irritated by all this, might just shrug and walk away. Thousands of people wake up every day, smile, and say: "I don't have to play tennis any more or listen to people who do." One day the same people might wake up and say: "I don't have to drink wine either, or listen to—or read—people who do."

SEPTEMBER, 1981

## FOR THE THANKSGIVING MEAL, AN ALL-AMERICAN CHOICE

We have always been a drinking nation; the rise and fall of Prohibition proved that. Even so, serving wine at the Thanksgiving table fifty years ago would have seemed somehow un-American, almost subversive.

In Norman Rockwell's classic *Saturday Evening Post* depiction of this holiday dinner, not a drop of alcohol is to be seen. But don't the ruddy cheeks of Rockwell's menfolk hint at an earlier pass or two at the communal jug? And Grandma's comely blush—is that from kitchen heat or the cooking wine?

Today, hundreds of thousands of us pour some kind of wine with the most American of all meals. The wine may not yet have the cachet of the creamed onions, but we've come a long way.

What wine to drink? Certainly it should be American, as Thanksgiving remains an American festival—unless the French appropriate it the way they have filched Halloween. And it should be a moderately priced wine; Thanksgiving is not an elitist celebration.

Sadly, moderately priced is easier said than done. In going after the more affluent market, the American wine industry has brushed off the customer of limited means. There is no lack of low-price wine, but most of it is mediocre. The lower end of the market has been left to the French, the Chileans, and the Italians.

Even so, there are some good American wines to be had at reasonable prices. I had hoped to come up with a list of good wines at $10 and under. I could do this only with chardonnays, and even in this category, raising the bar to $15 improved the picture markedly.

There are actually more white wines, even chardonnays, in the $10 range than there are reds. That is fortunate, because for many Americans, wine *is* chardonnay. My chardonnay list would include Estancia, Gallo Sonoma, Indigo Hills (another Gallo label), Meridian Santa Barbara, Kendall-Jackson, and two from Washington State—Columbia Crest and Hogue Cellars. All of them are 1998s and less than $10. I particularly liked a chardonnay called Morro Bay; it has good acid.

In the up-to-$15 group, I liked Rodney Strong Chalk Hill, Edna Valley, and Simi. They are worth the extra money. Don't expect Burgundian elegance from any of these wines, but they do offer interesting flavors and can stand up to almost any Thanksgiving meal.

This is mostly a search for red wines, however. When it comes to American reds, zinfandel is by far the most intelligent choice. First of all, it is the most American of the top varietals, even though, like most of us, it came from somewhere else. We are deluged with dull cabernets and merlots, one virtually indistinguishable from another, but somehow zinfandel seems to cling to its individuality. What's more, its bright spicy flavors make any meal taste good.

A lot of thin zinfandels were made in 1997, but in fact, some were quite good. Particularly attractive is a 1997 from Rancho Zabaco, yet another Gallo label. This wine, which includes 12 percent petite sirah, is made from Dry Creek Valley grapes and is $16. Dry Creek is one of California's premium zinfandel sources.

Gallo's own label, a zinfandel from the Barelli Creek ranch in the Alexander Valley, is outstanding. From the same appellation, Alexander Valley Vineyards' Sin Zin is a bit more smoothly textured and should appeal to people who like goofy names. Both are around $15.

The 1998 zinfandels, which overall are probably better than the 1997s, are also more expensive. Of the few I've tried, I can recommend Mother Clone from Pedroncelli Winery in the Dry Creek region in Sonoma. This is a good chunky wine with intensity. It's around $12. Seghesio Winery's Sonoma Zinfandel has good spice in the taste and bouquet. Seghesio has long produced fine zinfandels, and this is one of its latest at a truly reasonable price. These wines are in the $16 category.

The $10 cabernet sauvignon is still around, but it is in disguise. Lots of cabernets selling for $20 are the old $10 wines, unchanged except for the price. A few are more affordable. Beaulieu Vineyard's Rutherford, for example, is around $15, up from $9 a few years ago. But given the wild inflation of wine prices, it is still a good buy. Likewise for St. Francis, and it sells for a dollar or two less.

Both these wines are 1997s, as is the most recent release of R. H. Phillips's Toasted Head Merlot. R. H. Phillips is a winery in the northern Central Valley, in an area the Giguiere family, who own the place, call the Dunnigan Hills. This is an intriguing wine with touches of malbec, petit verdot, and cabernet sauvignon. R. H. Phillips also makes an Australian sort of blend with 55 percent Mendocino cabernet sauvignon and 45 percent of its own syrah. These two wines sell for around $15.

R. H. Philips has another merlot, into which it tosses two different petite sirahs, some petit verdot, and alicante bouschet. It is aged in oak for a year, then sold for around $9. At that price you can invite those cousins you haven't seen for years.

If you are still into merlot—not everyone has switched to syrah yet—consider one from Francis Ford Coppola's Diamond Series. This is not a wimpy merlot with no backbone. It comes from the same winery as Mr. Coppola's famous Rubicon.

If you're adventurous, there's Pelligrini's merlot from Long Island's North Fork. This is merlot in the St.-Émilion style. Elegant and delicious.

NOVEMBER, 2000

## OLD CHAMPAGNE NEEDS A LOVING HOME

After the repeal of Prohibition, the "21" Club had little use for its acclaimed secret cellar. With everyone drinking openly once again, the old hideaway where *tout New York* had whiled away the so-called dry years was turned into a wine *cave*.

When Mario Ricci, the famous sommelier of "21," first took me through the old speakeasy, its chief attraction, aside from nostalgia, was the wine that the club's best customers had stored there. Mr. Ricci was particularly proud of the Champagne that prominent fathers had laid down in anticipation of their daughters' weddings fifteen, twenty, or twenty-five years hence.

"But isn't most Champagne meant to be drunk soon after it's sold?" I asked.

"Yes," Mr. Ricci said, "but tell that to a new father."

The wine industry has done a thorough job of convincing its clientele that old is better than young. A lot of aged and very undrinkable wine is sold these days at very high prices. From time to time, old wines prove to be very good indeed, but far less often than gullible buyers think.

Champagne is a special case. It is good when it is young; it is good when it is old. In fact, it is almost two different wines. The Champagne houses insist that they make it to be drunk when it is released. Better than 95 percent of the 200 to 300 million bottles turned out every year are consumed within a week or two after their purchase. But Champagne definitely ages well. It's just not to everyone's taste or pocketbook when it does.

In 1998, a remarkable auction of older Champagnes took place in London. Among the wines sold were Pol Roger from 1892, 1914, 1921, and 1934; a 1907 from Heidsieck & Company; Krug from 1947, 1949, 1959, and 1961; and Veuve Clicquot from 1937 and 1947.

Tom Stevenson, a British wine writer, sampled many of the Champagnes before the sale. He concluded that the Pol Roger 1892 and 1914, the Heidsieck 1907, and a Moët & Chandon 1921 were four of the five greatest wines in the auction. A number of wines from the 1980s and 1970s were rejected as being past their peak. What's more, Mr. Stevenson added, each wine selected for auction was "in exceptional condition" and could still be "appreciated as Champagne, which means complete with fruit and fizz, no matter how old."

I've had my share of old Champagne—the 1911 and the 1914 Bollinger, with Christian Bizot when he was running that house, and various bottles from the 1920s and 1930s. But—and this is a big "but"—every one of those wines, like the wines in the London auction, came straight from the cellars of the Champagne houses that made them.

All true Champagne has at least some age before it's sold. Nonvintage Champagne must be aged for at least a year in the bottle before it is released for sale, and the bottling cannot take place until at least January 1 of the year following the harvest. Vintage Champagne cannot be sold until three years after the harvest. Actually, most big firms age their nonvintage Champagnes for two to three years and their vintage for three to five.

Special bottlings, like Bollinger's R.D., may be held in the cellars for a decade or more. R.D. stands for "recently disgorged." Disgorgement is one of the final steps in making Champagne, in which the sediment is removed from each bottle before it receives its *dosage,* or sweetening, and its final cork.

The sediment, or *lees,* is composed of dead cells of the yeast that originally caused the wine to ferment. The lees add to the wine's body and age. The 1985 R.D., released in 1997, was disgorged after eleven years on the lees; the 1981 was released late in 1999 after eighteen years on the lees.

These wines are available on special order. So are older Krugs and some Pol Rogers—not including the 1914 Pol Roger, which was disgorged in 1944, thirty years after it had first been bottled, and is apparently still in excellent condition.

All these wines have been aged by the people who made them, though. Enthusiasts with temperature-controlled cellars can age their own Champagne. Some Champagne fans believe that the long pre-disgorgement aging should be followed by a long post-disgorgement aging. A wine that is high in acidity, even after being disgorged, could probably do with another decade of aging, these Champagne fanatics insist.

Perhaps so, but keep in mind that these practices have nothing to do with that bottle of Dom Pérignon you stashed away in a hall closet five years ago. It might be okay; it could also be dead as a doornail. Champagne, real Champagne, is fine wine and must be cared for like any fine still wine.

Actually, drinking old Champagne is a connoisseur's game, and the Champagne makers are the last to encourage it. In the Champagne country east of Paris, they like their bubbles fresh and lively. With good reason: There is as much to say for the vivaciousness of a newly released good Champagne as there is for an old mellow one.

As an aperitif, which is the way most of us drink Champagne, its liveliness and elegance give a lift to the party, and its sharp acidity clears the palate for the meal to come. And compared with good Bordeaux and Burgundy, the price of young Champagne is quite reasonable. An extra bottle or two won't seriously lighten anyone's wallet.

APRIL, 2000

## BEAUJOLAIS PRESENTS ITS NOBLER PROFILE

Sometimes it's called real Beaujolais; sometimes Beaujolais for adults. What it means is "not Duboeuf."

So powerful a force are Duboeuf wines in the Beaujolais market, and so distinctive is their style, that most wine drinkers assume that all Beaujolais tastes like Georges Duboeuf's Beaujolais-Villages. And this assumption is not far from wrong.

Most Beaujolais is produced by shippers, or *négociants,* who buy from growers and cooperatives, and who turn out a refreshing, fruity wine that can be served chilled or not, one that goes with just about everything but the most elegant meals.

Some 35 million gallons of Beaujolais are produced each year. Half of that is simple Beaujolais; 25 percent is Beaujolais-Villages; and 25 percent is *crus.* About 55 percent is exported, some 12 percent to the United States.

The critic Stephen Tanzer, commenting on the Beaujolais market some years ago in *International Wine Cellar,* his newsletter, said that the wines had taken on "an overly perfumed, candied-fruit character," thanks to "widespread use of a commercial yeast strain originally popularized by Duboeuf but now in vogue throughout the region, as well as high yields and heavy chaptalization." (Chaptalization is the addition of sugar to the fermentation, not for sweetness but to raise the alcohol content of wines from what might be 10 percent or less to 12 percent or more.)

The immense popularity of Beaujolais, fueled by the nouveau Beaujolais craze, made it inevitable that serious wine drinkers would begin to look for something different. And it isn't difficult to find. Quantities are small, but the wines are out there.

Dozens, perhaps many dozens, of small producers are intent on pushing the gamay grape, from which all Beaujolais must be made, as far as it will go. The result is a fascinating selection of Beaujolais wines that are richer, darker, more intense, and drier than what most of us have come to expect from the Beaujolais region of southern Burgundy.

Many of these small producers continue to supply the big *négociants.* But then they hold back a portion of what they make to produce a more complex wine.

I discovered some of these wines, particularly those of Guy Breton and Jacky Janodet, in the early 1990s in Paris restaurants like the St.-Amarante, near the Bastille, and À la Courtille, in the 20th arrondissement, where small-batch Beaujolais were a feature. I say "were" because À la Courtille closed and the St.-Amarante was sold to a couple less committed to these wines.

Fortunately, many Paris bistros now offer this kind of Beaujolais. And it can be found in the United States if you know where to look. In New York, Al Hotchkin, at the Burgundy Wine Company in Greenwich Village, specializes in expensive Burgundies, but he has a soft spot for good Beaujolais.

Mr. Hotchkin offers small quantities of Beaujolais from Patrick and Elizabeth Giloux, Brouilly from Alain Michaud, Morgon from the Domaine Savoye, and Moulin-à-Vent from the Domaine du Granit. How small? The Gilouxes sell off their wines from younger vines. They hold back the wine from their forty-five-year-old vines, which rarely comes, even in good years, to more than six hundred cases, and not all of those travel to our shores. Alain Michaud is lucky to make a few hundred cases of his Cuvée Prestige Brouilly from ninety-year-old vines. Mr. Hotchkin advises laying down the 1998 vintage for five years. These are not mass-market wines.

Burgundy Wine Company even has wines from Georges Duboeuf. Though Mr. Duboeuf buys from four hundred growers, he also makes some special Beaujolais as well, often for top restaurants in his region like Georges Blanc, Paul Bocuse, and Léon de Lyons—and, most likely, for himself. Anyone who believes that Beaujolais doesn't age should try a couple of Duboeuf Moulin-à-Vents, the 1993 G.D.F. Cuvée or the aged-in-oak 1995 in magnum. I have had Beaujolais from the 1950s with Mr. Duboeuf.

More than half of all Beaujolais goes into simple Beaujolais and Beaujolais-Villages. But the region has areas where the wine is good enough and distinct enough to carry its own geographic name. These are the *cru* Beaujolais: Brouilly, Côtes de Brouilly, Chénas, Chiroubles, Fleurie, Juliénas, Régnié, St.-Amour, Morgon, and Moulin-à-Vent.

Most of the big Beaujolais shippers are in the Beaujolais region, but most of the prominent Burgundy shippers, like Louis Latour, Louis Jadot, and Joseph Drouhin, also sell Beaujolais. In 1966 Jadot bought a sixty-six-acre estate in Moulin-à-Vent called Château de Jacques, and he is now offering five Moulin-à-Vents from separate vineyard parcels within the estate. These are traditional Beaujolais, full and rich, aged in oak and capable of improving even more with bottle age.

Because these wines are imported in small quantities, it is not always easy to learn which are available in this country. But a few great traditional Beaujolais should be snapped up if you find them anywhere: Fleurie, from Yves Métras; Morgon, from Jean Foillard; and Moulin-à-Vent, from Paul Janin.

What makes these special Beaujolais attractive is the same thing that has always made Beaujolais attractive: the price. Given the insane prices of so many wines right now, Beaujolais and the delicious wines pouring in from southern Italy and Sicily keep many wine drinkers from switching to iced tea. Even now, fresh, fruity Beaujolais is widely available at less than $10, and most of these old-style small-batch *cru* Beaujolais still sell for $15 to $20 a bottle. The relationship between price and quality is hard to match in wine from any other region.

MAY, 2000

## CABERNET ON TAP
## (NOT FOR CORK SNIFFERS)

There's no wine in the house. Run out and get a box of chardonnay."

A box of chardonnay? Sure, why not? Chardonnay, cabernet sauvignon, sauvignon blanc, whatever you like.

While you may have been poring over vintage charts, joining tasting groups, trying to use precisely the right glass, and wondering if your precious 1982s are over the hill, lots of people have been just drinking wine.

There is no law that says you have to be a connoisseur to enjoy a glass of wine. Nor is it written that you have to read a wine textbook before you pull a cork—or, more often, unscrew a cap.

Don't get nervous: this is not going to be an anti-intellectual tract or an exercise in praise of wine boorishness. It's just pleasant, now and then, to recognize that there is another world of wine, far removed from the precious confines of the cork sniffers.

Who are the cork sniffers? The term is one of mild derision applied to folks who take wine very seriously. Who applies it? Beer salesmen, whiskey salesmen—people in the more traditional, more unbuttoned segments of the liquor business.

All cork sniffers drink wine, presumably, but not all wine drinkers are cork sniffers. What provokes this insight is a report, admittedly self-serving, from the people at Franzia Vineyards in California predicting that in 1993, Americans will drink some 40 million gallons of wine out of boxes.

Now, 40 million gallons is a lot of wine, more than 16.5 million cases. No cork sniffer would be caught dead with a box of wine, but if the Franzias are to be believed, there are plenty of Americans who think the wine box is a good idea.

The wine box, or as it is more commonly known, the bag-in-the-box, has been around for at least thirty years. It first enjoyed considerable popularity in Australia. It has had some success in Europe but until recently never made any serious inroads on the American market. In this country, about the only thing shared by connoisseurs and the consumers of inexpensive wine has been the conviction that wine should come in bottles.

There is little doubt among wine professionals that the recession has given the bag-in-the-box a boost. Indeed, there has been a minor revolution in the wine business as a huge segment of the market discovered that wine does not have to be expensive to taste good.

Franzia has about 90 percent of the wine-box business in this country, mostly under its own name but also under the Summit and Colony labels. Almaden Vineyards, now a subsidiary of Heublein, produces some wine boxes. Heublein also sells some under the once legendary Inglenook label.

The most common size of wine box bought by individuals is five liters, which holds slightly less than seven bottles of wine. (There is also an eighteen-liter box, bought primarily by restaurants.) The box—Franzia calls its version the Winetap—takes up about a third less space than a four-liter glass jug and about the same space as four single bottles. A built-in tap makes pouring wine simple; there is no need to remove the box from the refrigerator. The bag that holds the wine inside the box is an airtight polyethylene container that collapses as the wine is poured. No air gets in, which means the wine can last for days.

What about quality? Within their category, box wines are first-rate. If that sounds like a qualified statement, it is. They are relatively soft and fruity with a touch of sweetness. A devotee of great Bordeaux will find little excitement here. But devotees of great Bordeaux are few, and people who like an occasional glass of easy-to-drink, unpedigreed wine are, apparently, legion.

In the past, bag-in-the-box wines were almost exclusively generics, that is, blends of wine from various grapes, given names like burgundy and claret. Many box wines are still generics, like Franzia's Mountain Chablis, which is certainly not Chablis and probably has nothing to do with mountains but is a fresh, pleasant wine. Burgundy and Rhine are the other two generics in the line.

But now there are varietals in boxes. A varietal is a wine that bears the name of the principal grape from which it is made; cabernet sauvignon, for example, or chardonnay, or the most popular of all, white zinfandel. Other varietals found in boxes include sauvignon blanc, white grenache, French colombard, and chenin blanc. Note that only two of the eleven Franzia varietals are red wines: cabernet sauvignon and Burgundy.

Not too many years ago, the idea of putting cabernet or chardonnay in a box would have been unthinkable. They were the noble grapes of France and of the great vineyards of northern California. They were hard to grow and expensive to turn into wine.

Today there are plenty of good grapes available, and wineries like E. & J. Gallo, Glen Ellen, Fetzer, and Sutter Home produce millions of bottles of these once rare wines at highly competitive prices.

There is little argument over the fact that the bag-in-the-box is one of the most economical and efficient ways to package wines that will be drunk soon after purchase. Image is what has held the boxes back. Wine drinking has long been inextricably intertwined with prestige and tradition. What other product could survive with so inconvenient a package as the cork-stoppered bottle?

Finally, faced with a shortage of good corks and a proliferation of bad ones, the industry is grudgingly turning to alternative stoppers. And there is less consumer rejection that anyone predicted. In France, for the longest time, the only wines found in boxes were heavy reds and dull whites from the furnacelike vineyards of Languedoc-Roussillon. In recent years, good Côtes du Rhône has made its appearance in a box, along with some pleasant whites from the Loire. Can Beaujolais be far behind?

And in this country, will wine in a box move up the social scale? With chardonnay and cabernet already boxed in, it looks like it already has.

AUGUST, 1993

## TO SEND IT BACK OR NOT?

The phone call came from Jean-Claude Baker, who runs Chez Josephine, a bistro on 42nd Street in New York. A customer had ordered a $100 bottle of wine, then announced that it was "no good" and demanded his money back. Mr. Baker gave the fellow his money even while convinced that there was nothing wrong with the wine.

What, he wanted to know, should one do in that kind of situation?

It's somewhat comforting to hear that even veteran restaurateurs get squirmy when the subject is returning a bottle of wine. Wine is intimidating to many diners and, most of the time, when they think a bottle is bad, they lack the confidence to complain and end up drinking something that should be poured down the sink. On the other side of the equation are the wine boors, who know too little to accurately judge a wine's quality but are only too eager to pronounce it inferior.

(Henri Soulé, who ruled over Le Pavillon restaurant in New York in the 1950s, dealt summarily with the boors: when someone complained about a perfectly good bottle, the captain would summon Mr. Soulé. "Of course we will replace it," he would say, adding that he found nothing wrong with the wine. The captain would then say he had found nothing wrong with it, either. Another captain might be consulted; he, too, would find nothing wrong. By then the cocky customer would be looking for a table to crawl under.)

The send-back problem has been compounded in recent years by the huge rise in the price of wine at restaurants, and the increasing number of "serious" wine drinkers. People who have little or no idea of what a great wine should taste like assume that anything they pay $300 a bottle for is going to be delicious. And it may well be—to someone who has been drinking wine for years and knows what to look for in a twenty-year-old Bordeaux.

A newcomer to music is not expected to rhapsodize over the Bach B minor mass. So why should a wine neophyte be expected to find joy in an austere St.-Estèphe from 1970? It's the job of a good sommelier to see problems brewing and head them off by suggesting, delicately, something more appropriate. If the customer persists, he can't complain afterward—assuming there's nothing actually wrong with the wine.

If a wine is in fact bad—if it's corked, for example—a good waiter will replace it without question. In turn, most wholesalers who supply restaurant wines will take the bottle back and not charge for it. But if it's an older wine, one the restaurant may have had in its cellar for a long time, it's hard to blame the distributor, or even the winemaker. Some restaurants have a policy of not replacing wines twenty years old or more, and they let the customer know this when they order them.

Corkiness is the most common wine problem. The cause is a mold that thrives in the bark of the cork oak tree. Proper treatment during the cork-making process eliminates it, but affected corks can and do slip through. They give the wine the smell of a moldy, damp basement and impair the taste as well. Corkiness is found in both young and old wines, which is why some wineries now use only plastic corks. One small California winery, PlumpJack, made news recently by putting a screw-top cap on a $135 bottle of cabernet.

Unfortunately, few restaurant customers know how to identify corkiness.

Good sommeliers understand. "Too few bottles are sent back," said Larry Stone, the wine director of Rubicon, a San Francisco restaurant. "If you look at the statistics on corked wine, you know there should be more returns."

In the mid-1990s, the problem of corked wines was serious, with some distributors and retailers claiming to find a corked bottle in almost every other twelve-bottle case. Faced with mass defections to plastic corks, the cork manufacturers improved quality considerably. But even if only one percent is contaminated, a busy restaurant can expect to serve one or two bad bottles a night. "What worries me," said Darryl Beeson, wine director at the Mansion on Turtle Creek in Dallas, "is the ones we don't catch."

The easiest way for a diner to handle the problem is to ask the waiter or wine steward's advice. If he or she thinks the wine is "off," it will be replaced. Asking the server's advice is flattering and non-confrontational. In a good restaurant it almost always works. In places where wine is not often served or where the staff has no wine training, there may be a problem.

How does one detect corkiness? If the wine has a moldy smell, even a slight one, it's probably corked. A friend of mine learned to recognize it by asking waiters to let him smell any corked bottles they may have retrieved that day.

Other wine problems are easier to deal with. If the wine is at the wrong temperature—a red wine too warm or a white wine too cold—a couple of minutes in an ice bucket for the red and ten minutes out of the bucket for the white should do the trick. Wine that has been aged prematurely, usually because of a dried-out cork, should be returned. Older wines, even great ones, will often throw a sediment. It has no effect on the wine and can be eliminated by decanting.

Tartrate crystals, the clear granules occasionally found on the bottom of the cork in a bottle of white wine, are also harmless. European wine drinkers shrug them off, but because they make squeaky clean Americans nervous, producers strive to eliminate them before the wines go to market.

Then, let's face it, there is the wine that, for one reason or another, has turned completely bad. It will smell like vinegar and you don't have to go to school to recognize that.

AUGUST, 2000

# GLOSSARY

ALCOHOL. In wine, a colorless spirit formed by the action of enzymes in yeast cells feeding on sugar in the grape juice. The process, called fermentation, produces almost equal amounts of ethanol, another name for eythyl alcohol, and carbon dioxide gas.

*APPELLATION CONTRÔLÉE*. A French term meaning a specific place where a certain type of grape can be grown. Appellation laws determine the amount of grapes that can be grown and the amount of wine that can be made from them.

AROMA. The smell associated with a young wine. It fades as the wine matures and is replaced by bouquet.

*ASSEMBLAGE*. A term associated particularly with Bordeaux, it refers to the "assembly" of a chateau's wine from the most recent vintage by blending the various lots of wine that were picked and fermented separately.

BARREL CELLAR. The room or space where wine is matured in wooden casks, usually oak.

BODEGA. A winery in Spain.

BODY. In wine tasting a term that implies weight or the impression of weight; the opposite of thinness. A wine with lots of body can be awkward and clumsy if the weight is not balanced by finesse.

BOUQUET. The delightful smell of a well-matured wine.

BRUT. A French term designating the driest Champagnes. A brut Champagne usually has less than one percent sugar. Champagnes with fanciful designations like *brut sauvage* may have no sugar whatsoever. Curiously, extradry Champagne is sweeter than brut.

CELLARER. If a cellar in wine terms is a place for storing wine, the cellarer is the man or woman in charge of the cellar, keeping temperature and humidity at proper levels and topping up casks in which some wine has evaporated.

**CHAIS**. French word for the part of the chateau or winery where the wine is made.

**CHAPTALIZATION**. The process of adding sugar to grape juice at the time of fermentation, not for sweetness but for alcohol. Banned in some wine regions with abundant sun and lots of natural sugar in the grape juice. Named for Napoleon's minister of agriculture, who is said to have urged the practice. Actually a similar process, involving honey, had been used since Roman times.

**CHEWY**. Another wine-tasting term applied to wines that are tannic and relatively rough in the mouth.

**CLONE**. In wine, a vine that has been reproduced asexually and retains the same generic characteristics as the parent. Clonal selection is an important tool in creating vines with special qualities and for specific growing regions.

**CORKED**. A wine bottled under a defective cork that has taken on the cork's bad odor and taste. Corked wine usually tastes and smells of mold. The trend to plastic corks in recent years is a reaction to the problem of defective genuine corks.

**CRU**. A French term meaning "growth" in English but a specific vineyard when used in France. Thus a *grand cru* in Burgundy means a vineyard of the highest level. A *cru classé* in Bordeaux refers to the vineyard or the wine of a chateau ranked in one of the various classifications, especially the famous classification of 1855. The best wines in that ranking are still known as *premiers crus,* or first growths. A *cru bourgeois* is a wine from a Bordeaux chateau ranked just below the *cru classés*. There are some 250 *cru bourgeois* chateaux.

**CRUSH**. An American term synonymous with the harvest of wine grapes. Most grapes are picked for fruit and raisins. Only grapes for juice and wine are crushed.

**DEMI-SEC**. French for half dry, it refers to Champagne that it quite sweet, containing up to 5 percent sugar.

**DISGORGEMENT**. An important step in making most sparkling wine and all Champagne. It involves the removal of sediment for each bottle and replacing it, first with the dose of sweetened wine called the dosage and then the final cork.

**ENOLOGY**. The scientific study of wine and winemaking. The name derives from Oeneus, a Greek wine god.

**FAT.** Yet another wine taster's term. It generally refers to wine that has body but little tannin or structure.

**FERMENTATION.** The process by which grape juice becomes wine when the enzymes in certain yeasts attack the sugar in grape juice and transform it into alcohol and carbon dioxide. Louis Pasteur was the first to show how the process works.

**FINISH.** Another wine-tasting term, it refers to the length of time the taste of a wine lingers in the mouth. If the taste lasts for a while, the wine is said to have a long finish.

**LEES.** Sediment deposited in a young wine from the barrel or tank in which it is stored after fermentation. Also the dead yeast cells left in some wines, like Muscadet, after fermentation, to give the wine more body.

**LEGS.** When a wine is swirled in a glass, some of it clings to the sides and descends slowly in a pattern called legs. It is mostly alcohol and has nothing to do with the quality of the wine. In Germany the "legs" are called *Kirchenfenster,* or church windows, because the wine seems to form a gothic arch.

**NÉGOCIANT.** French term for a wine merchant. Originally, *négociants* confined themselves to buying wine from the growers, bottling, aging, and selling it. Now most négociants, particularly in Burgundy, are themselves major vineyard owners.

**NOSE.** Wine tasters call the aroma and/or bouquet of a wine its nose.

**OAK, OAKY.** Oak barrels impart a vanillin-like taste to wines aged in them. A wine with too much of that taste is called oaky. Americans love the taste of oak, and various artificial means are used to give even unaged wines an oaky taste.

**pH.** A sign that shows the level of acidity in a wine. The lower the pH, the higher the wine's acid content. Acid is what gives a wine its freshness when young and is also what helps it age gracefully.

**PHYLLOXERA.** A microscopic insect native to the eastern United States that was introduced into France around 1860 for laboratory experiments. Within twenty years it wiped out almost all the wine-producing vineyards of the world. It was finally controlled when European vines were grafted onto resistant American roots. A second epidemic did millions of dollars' worth of damage in California in the early 1900s when a mutant version of the bug destroyed thousands of acres of vines thought to be phylloxera-resistant.

**PREMIER CRU**. See cru.

**ROUNDNESS**. Taster's term for a quality in a wine in which all the elements—fruit, acid, and tannins—are in balance.

**STRUCTURE**. A term for the quality tannins impart to an otherwise soft red wine. Cabernet sauvignon, often a tannic wine, may be added to a softer merlot to give it "structure."

**TANNINS**. Organic compounds found in the bark and stems of many trees and plants. They are also found in the skins, seeds, and stems of wine grapes. The longer the grape juice is left in contact with the skins and stems, the more tannins the wine will have. It must be balanced with the proper amount of fruit and acid to produce a superior wine.

**TONNELIER**. In French, a barrel maker. His shop is known as a *tonnelerie*.

**VARIETAL**. A wine named for the grape from which it is made. Americans are more familiar with varietal names than place names. A pinot noir in America will be a red Burgundy in France. Most Tuscan red wines are made from the sangiovese grape but are not known by that name in Italy. Most white Burgundies are made from chardonnay grapes, but the name never appears on the label. In this country we often prefer to know what the grape is rather than where the wine comes from.

**VENDANGE**. Harvest, in French.

**VENDEMMIA**. Harvest, in Italian.

**VERTICAL**. A wine tasting of several vintages of a single wine. A tasting of ten vintages of Château Lafite-Rothschild would be a vertical tasting. A tasting of ten different Bordeaux from the 1961 vintage would be a horizontal tasting.

**VIN GLACIER**. Ice wine, or *eiswein*. Wine made from grapes that have been left on the vine into deep winter. When they freeze, the water in them is separated from the juice and acid, which can have double the intensity of normal wines. In recent years, some winemakers have experimented with freezing normal grapes in the winery. This technique is also used to turn raspberries and other fruits into interesting ice wines.

**VIN LIQUOREUX**. Unctuous, sweet white wines such as Sauternes made from late-harvested grapes that have retained high amounts of natural sugar.

**VINHO VERDE**. Literally, "green wine" in Portuguese. One-dimensional red and white wines from the northwest corner of Portugal. The acidic, fresh whites are preferable, particularly those made from the Alvarinho grape, which is identical to the albarino wines made just over the Minho river, in the Spanish province of Galicia.

VINIFICATION. Just another term for making wines. One "vinifies" the grapes.

VITICULTURE. In the simplest terms, growing grapes. For many years, Americans were convinced wine was made in the winery, as opposed to Europeans, who were convinced that the best wine came from the best vineyards. Hence their preference for place names for their wines. American winemakers have at last come forward to this point of view. The grape grower has come into his own.

# INDEX